STARRY NIGHT COLLEGE WORK

to accompany

21st Century Astronomy

FOURTH EDITION

STARRY NIGHT COLLEGE WORKBOOK

to accompany
21st Century Astronomy

FOURTH EDITION

*Laura Kay, Stacy Palen, Brad Smith, and
George Blumenthal*

Steven Desch

GUILFORD TECHNICAL COMMUNITY COLLEGE

Donald Terndrup

THE OHIO STATE UNIVERSITY

W · W · NORTON & COMPANY · NEW YORK · LONDON

Associate Editor: Jennifer Barnhardt
Project management and composition by Westchester Publishing Services
Production manager: Eric Pier-Hocking

Printed in the United States of America

ISBN: 978-0-393-92181-6 (pbk.)

W. W. Norton & Company, Inc., 500 Fifth Avenue, New York, N.Y. 10110
wwnorton.com

W. W. Norton & Company Ltd., Castle House, 75/76 Wells Street, London W1T 3QT

1 2 3 4 5 6 7 8 9 0

CONTENTS

PREFACE

Astronomy is a challenging subject, but a rewarding one. The challenges come in many forms. Because astronomers have learned so much about planets, stars, galaxies, and the universe, there is a lot of material to cover in any college course. Students have many new concepts to learn, and the vocabulary of astrophysics is often strange. Finally, many of the key ideas are hard to visualize because our lives do not give us direct experience of orbits, atomic nuclei, quasars, and the like.

The exercises in this workbook are designed to help you become more familiar with many astronomical phenomena. We begin by thoroughly exploring the appearance of the sky and the motion of the Sun and Moon as seen from Earth. Later, we step out into the Solar System and explore the orbits of the planets and other objects. We then use Starry Night to gather information on the basic properties of stars. Finally, we leave the confines of the Milky Way galaxy to look out into deep space.

The goal of these exercises is to help you understand many key concepts from the text. After completing the assignments you will be skilled in operating Starry Night and can learn much more on your own.

Please note that the instructions in this manual were written for PC users. Following are a few differences between the controls on a PC and those on a Macintosh:

- Right-click on the PC corresponds to CTRL-click on the Macintosh.
- On the PC, the object lists in the Info and other windows are expanded by hitting the + or − buttons next to the list name. These are called "Expand layer" and "Collapse layer." On the Mac these look like a right-arrow and down-arrow, respectively.

A complete guide to the controls can be found by using the Help/User's Guide menu. This displays a PDF file. The Quick Tips on page ix of this booklet are meant to summarize the items in the Getting Started and Basics section of the User's Guide.

QUICK TIPS

- **Orient your gaze toward a compass direction.**
  Use the N, S, E, or W buttons located toward the upper right under the GAZE display. Holding the space bar while clicking one of these buttons will make the gaze shift instantly.
- **Change your gaze upward, downward, left, or right.**
  Use the arrow keys on your keyboard. Alternatively, move the cursor in the display until it looks like a hand; click and drag to change the gaze direction.
- **Zoom in or out toward the center of the display.**
  Use the + or − buttons, located at the upper right under the ZOOM (WIDTH × HEIGHT) display.
- **Zoom in or out toward an object on the sky.**
  Move the cursor toward the object, then use the mouse wheel to zoom in or out.
- **Stop the flow of time.**
  Click the STOP TIME button (■).
- **Start the flow of time.**
  Click the RUN TIME FORWARD button (▶). The time step between frames is shown in the TIME FLOW RATE display just above this button.
- **Run the flow of time backward.**
  Click the RUN TIME BACKWARD button (◀).
- **Advance the time forward by one time step.**
  Click the STEP TIME FORWARD button (▮▶).
- **Move the time backward by one time step.**
  Click the STEP TIME BACKWARD button (◀▮).
- **Multiply the time flow rate.**
  Click on the down-arrow (▼) next to the TIME FLOW RATE display. Select the multiplier from the top part of the dropdown window. For example, to advance at 300 times the normal rate, select 300×.
- **Set a specific time flow rate.**
  First, select the units for the time flow rate by clicking on the down-arrow (▼) next to the TIME FLOW RATE display. Select the desired units from the dropdown menu. Then click on the number in the TIME FLOW RATE display and type the rate you want. For example, to set the rate to five minutes, first select minutes from the dropdown menu, then click on the current time step and type **5**.
- **Set a specific date.**
  Move your cursor to the TIME AND DATE display and click on the item you want to change. Either type in the new value or use your keyboard up- and down-arrows to select one. The right- and left-arrows are used to move to the next or previous item in the TIME AND DATE display.
- **Open/close a tab for more options.**
  Click on one of the options tabs to the left (FIND, OPTIONS, FAVORITES, etc.). Click on the name again to close.
- **Identify a particular object.**
  Move the cursor over the object until it changes to indicate the heads-up display (▤). Left-click when the cursor is this shape, and a label will appear on the display with the name of the object. A right-click will show a menu that can be used to display information about the object.

STUDENT EXERCISE | The Celestial Sphere

GOAL

- To investigate the apparent daily motion of the sky caused by Earth's rotation

READING

- Section 2.2 – Earth Spins on Its Axis

We will see what the daily motion of the sky looks like from Earth's North and South Poles. This exercise will animate the information shown in Figure 2.7 in the textbook.

It is convenient to imagine the sky as a large sphere centered on the Earth. This is called the **celestial sphere**. In reality, the stars and planets are all located at different distances from Earth.

Approximately every 24 hours, the sky appears to rotate around two points. These are called the celestial poles. There are two poles, the **north celestial pole** and the **south celestial pole**. At the North Pole, the north celestial pole is at the **zenith** – the direction straight up.

From either pole, an observer will see one half of the celestial sphere. The half visible from the North Pole is called the northern celestial hemisphere. An observer at the South Pole will see the southern celestial hemisphere. At the poles, all visible stars are above the horizon continuously; they neither rise nor set. Such stars are called **circumpolar**.

To navigate the sky, we need to define directions. From any point on the sky, north is the direction from that point toward the north celestial pole. West is the direction of the apparent motion of the celestial sphere (i.e., the stars are constantly moving toward the west). East is the opposite direction from west.

Astronomers divide the sky into constellations, which are areas on the sky around particular groups of stars. The stars in each constellation, in general, are not physically associated but instead simply happen to lie in the same direction. Constellations are further discussed in section 2.3; here we use them to help visualize the daily motion of the sky.

SETUP

- Start Starry Night.
- Stop the flow of time with the STOP TIME button (■).
- Using the dropdown menus, select OPTIONS / VIEWING LOCATION. A box labeled VIEWING LOCATION will appear.
  - Select the LATITUDE / LONGITUDE tab.
  - In the upper dialog box (LATITUDE), erase the current coordinate and type **90 N**.
  - Click the GO TO LOCATION button at the lower right of the dialog box.
  - Your viewing location will now shift to the North Pole. You can speed up the process by hitting the space bar while the motion is taking place. At this point, you may see the sky or you may be looking at the ground.
  - Click on the S button toward the upper right to point your gaze toward the south.
- Move the cursor over the window until it becomes a hand. If the horizon does not appear toward the bottom of the window, use the up-arrow key on your keyboard until the horizon appears toward the bottom of the window. Alternatively, you can use the handgrab tool to alter the direction of your gaze; click and drag upward until the horizon appears at the bottom of the window.

- Open the OPTIONS tab to the left by clicking on the word OPTIONS. Set up the following items by checking the appropriate boxes:
  - GUIDES: CELESTIAL GUIDES: POLES on, other options off.
  - LOCAL VIEW: DAYLIGHT off, LOCAL HORIZON on, other options off.
  - SOLAR SYSTEM: All options off.
  - STARS: STARS on, MILKY WAY off.
  - CONSTELLATIONS: BOUNDARIES on, LABELS on, STICK FIGURES on, other options off.
- Close the OPTIONS tab by clicking the word OPTIONS on the tabs.

ACTIVITY 1 – DIRECTIONS ON THE SKY

- Set the time flow rate to two minutes. To do this, click on the little arrow to the right of the TIME FLOW RATE box and select minutes from the dropdown menu. Then click on the number in the TIME FLOW RATE box and type a 2.
- Click the ▶ button to start the flow of time. Verify that the time displayed in the upper left is advancing. Let the motion run for a while, then stop the flow of time with the ■ button.

1. At the location of any star, west is defined as the direction of the apparent motion and east is the opposite direction. Is west toward the left or toward the right? What about east?
2. Do the stars seem to move parallel to the horizon or at a large angle to the horizon?

ACTIVITY 2 – DIRECTION OF ROTATION

- Move the cursor over the main window and click once. Using the up-arrow key on your computer's keyboard or the handgrab tool, raise the angle of your gaze upward until the north celestial pole is in the middle of the window. You are now looking toward the zenith, and the horizon should be out of view.
- Click on the ▶ button to start the flow of time. Let it run for a while, then stop the action.

3. Is the apparent rotation of the sky clockwise or counterclockwise?
4. In which constellation is the north celestial pole located?

ACTIVITY 3 – VIEW FROM THE SOUTH POLE

- Following the procedures described in the Setup, set the viewing location to the South Pole, which is at latitude 90 S. Point your gaze to the north using the N button. Use the up-arrow on the keyboard or the handgrab tool to place the south celestial pole near the middle of the window. Start the flow of time.

5. Is the apparent rotation of the sky from the South Pole clockwise or counterclockwise?
6. In which constellation is the south celestial pole located?
7. Do stars appear to move in the same direction at the North and South Poles? Explain why or why not.

Name: _____

Class/Section: _____

Starry Night Student Exercise – Answer Sheet
The Celestial Sphere

1. At the location of any star, west is defined as the direction of the apparent motion and east is the opposite direction. Is west toward the left or toward the right? What about east?

   _____

2. Do the stars seem to move parallel to the horizon or at a large angle to the horizon?

   _____

3. Is the apparent rotation of the sky clockwise or counterclockwise?

   _____

4. In which constellation is the north celestial pole located?

   _____

5. Is the apparent rotation of the sky from the South Pole clockwise or counterclockwise?

   _____

6. In which constellation is the south celestial pole located?

   _____

7. Do stars appear to move in the same direction at the North and South Poles? Explain why or why not.

   _____

   _____

Earth's Rotation Period

GOALS

- To investigate the apparent motion of the sky from an intermediate latitude between the equator and the pole
- To determine the length of the day as measured by the stars and the Sun

READING

- Section 2.2 – Earth Spins on Its Axis

This exercise will animate the information shown in Figure 2.12 in the textbook. We will view the motion of the celestial sphere as seen from a place between the equator and the North Pole or the South Pole. From these locations, certain stars will rise and set, while others will remain above the horizon at all times. The latter are termed **circumpolar**.

We also will determine Earth's rotation period, which is the amount of time it takes for Earth to spin once on its axis compared to the distant stars. We do this by noting the times that a particular star crosses the **meridian**. The meridian is a line that runs between the north and south points on the horizon through the **zenith** (the point straight up). The meridian divides the sky into east and west.

Earth's rotation period is called the **sidereal** day. The word *sidereal* refers to the stars. In many other exercises, you will be asked to observe the sky at intervals of one sidereal day or many sidereal days. If you do this, the stars will be in the same position at each observation.

We keep track of time using the Sun, which (on average) crosses the meridian every 24 hours. Intervals of 24 hours are called a **solar day** (or simply a day). In this exercise, we will see that the sidereal day is not 24 hours! The sidereal and solar days differ because Earth orbits the Sun. From Earth, it looks like the Sun moves eastward a little bit compared to the stars each day. The Sun's annual path on the celestial sphere is called the **ecliptic**. The motion of the Sun along the ecliptic is explored in another exercise.

SETUP

- Start Starry Night.
- Stop the flow of time with the STOP TIME button (■).
- Using the dropdown menus, select OPTIONS / VIEWING LOCATION. A box labeled VIEWING LOCATION will appear.
  - You can select your location from a list of places. To do this, select the LIST tab. Scroll up and down to select a location. Pick the location closest to where you live. Click on that location to select.
  - Alternatively, select the MAP tab. Click on a location on the map of the Earth.
  - Click the GO TO LOCATION button toward the lower right of the VIEWING LOCATION box.
- Click on the N button toward the upper right to point your gaze toward the north point of the horizon. (Students in the Southern Hemisphere should click the S button.)
- Open the OPTIONS tab to the left. Set up the following items by checking the appropriate boxes:
  - GUIDES: CELESTIAL GUIDES: POLES on, other options off.

- LOCAL VIEW: DAYLIGHT off, LOCAL HORIZON on, other options off.
  - SOLAR SYSTEM: All options off.
  - STARS: STARS on, MILKY WAY off.
  - CONSTELLATIONS: BOUNDARIES, LABELS, STICK FIGURES on, other options off.
- Close the OPTIONS tab by clicking the word OPTIONS.
- Move the cursor over the window until it becomes a hand. Click and drag the view until the horizon appears toward the bottom of the window. Alternatively, click once on the window and use the down-arrow on your keyboard to bring the horizon into view. Unless you have a very small window, or are observing from a high northern or southern latitude, you should see both the horizon and the celestial pole.

ACTIVITY 1 – CIRCUMPOLAR CONSTELLATIONS

- Set the TIME FLOW RATE to two minutes. To do this, click on the little arrow to the right of the TIME FLOW RATE box and select minutes from the dropdown menu. Then click on the number in the TIME FLOW RATE box and type a **2**.
- Start the flow of time, and let it run for about 24 hours. Because the option DAYLIGHT is off, you can see the stars even during the daytime. Stop the flow of time.

1. Make a list of constellations that are mostly or entirely circumpolar from your viewing location.

ACTIVITY 2 – RISING AND SETTING CONSTELLATIONS

- Use the E button to change your direction so you are facing east.
- Start the flow of time, and let it run for a bit, then stop. You should see stars rise on the eastern horizon.
- Pick a constellation that is rising in the east. Use the STEP TIME BACKWARD button (◀|) to move the time back to the point when the first stars of that constellation are rising. You can pick the bright stars that are connected by the stick figures. Note the time from the TIME AND DATE display at the upper left.
- Then use the STEP TIME FORWARD button (|▶) to advance the time until the last stars of that constellation are rising. Note the time.

2. What constellation did you pick? How long does it take the whole constellation to rise, in hours?

ACTIVITY 3 – EARTH'S ROTATION PERIOD

- Point toward the southern horizon with the S button. (Students in the Southern Hemisphere should click N instead.)
- Click the OPTIONS tab to expand.
- Under GUIDES, ALT-AZ GUIDES, select MERIDIAN. You will see a line appear running upward from the direction marker on the horizon.
- Close the OPTIONS tab by clicking on the word OPTIONS.
- Set the time to be noon (12:00:00 P.M.). Do this by clicking on the hours, displayed in the TIME AND DATE subwindow, and either typing in the hours you want (12) or using the up- or down-arrows to select the time. Do the same for the minutes and seconds, and also for A.M./P.M.
- Find a bright star that is near the meridian. Move the cursor to the star. When the cursor changes to the heads-up display shape (▤), click on the star. The name of the star will appear.
- Set the TIME FLOW RATE to five seconds.
- Zoom in so that the area near the star and the meridian are magnified. To zoom, place the cursor near the star you have identified and use the mousewheel to zoom in or out. (If you do not have a mousewheel, use the + and – keys in the upper right to zoom.) You may need to use the handgrab tool to have the star and the meridian both visible near the center of the screen.

3. Use RUN TIME FORWARD (▶) or RUN TIME BACKWARD (◀), along with STOP TIME (■), and determine the time when the star is on the meridian. You can use the STEP TIME FORWARD (|▶) or STEP TIME BACKWARD (◀|) buttons to place the star right on the meridian. Write down the time when the star is on the meridian.

   Now advance the time by 24 hours by clicking on the day value in the TIME AND DATE window and then hitting the up-arrow on your keyboard. (For example, you would change the **5** in the date October **5**, 2007.) Alternatively, you can click on the day value and type in a date that is one date later than the current one. Use STEP TIME FORWARD or STEP TIME BACKWARD until the star is on the meridian again. Write down the time and the interval between this meridian crossing and the previous one.

   Do this for several days, filling in the following table. (The first couple of entries are shown as an example only. Your time will vary depending upon your viewing location.) Do this for five days, then compute the average interval.

Date	Time	Interval
March 17, 2007	12:26:38 P.M.	
March 18, 2007	12:22:43 P.M.	23 hr, 56 min, 5 sec
. . . etc.	. . . etc.	
	Average	

4. What is the average interval of time between successive crossings of the meridian? This time is Earth's rotation period (1 sidereal day).
5. On average, how many minutes earlier do the stars cross the meridian each day?
6. For a star that crosses the meridian at about midnight tonight, what time will it cross tomorrow night? How about a month (30 days later)? What does this imply about the visibility of a particular star over the course of a year?

Name: _____

Class/Section: _____

Starry Night Student Exercise – Answer Sheet
Earth's Rotation Period

1. Make a list of constellations that are mostly or entirely circumpolar from your viewing location.

   _____

   _____

2. What constellation did you pick? How long does it take the whole constellation to rise, in hours?

   _____

3. Table of meridian crossings for the star:

Date	Time	Interval
	Average	

4. What is the average interval of time between successive crossings of the meridian? This time is Earth's rotation period (1 sidereal day).

   _____

5. On average, how many minutes earlier do stars cross the meridian each day?

   _____

6. For a star that crosses the meridian at about midnight tonight, what time will it cross tomorrow night? How about a month (30 days later)? What does this imply about the visibility of a particular star over the course of a year?

   _____

   _____

Motion of the Sun along the Ecliptic

GOAL

- To investigate the apparent motion of the Sun during the year caused by the orbit of Earth about the Sun

READING

- Section 2.1 – A View from Long Ago
- Section 2.3 – Revolution about the Sun Leads to Changes during the Year

As Earth orbits the Sun during the year, the Sun will be located in front of various constellations. This is illustrated in Figure 2.14 of the textbook. In popular astrology, it is said that the Sun "enters Gemini" or "enters Virgo." Many calendars will show the dates the Sun enters each constellation. Astrology columns in the paper show the full range of astrological dates that the Sun is in each constellation (e.g., Gemini: May 21 – June 20).

Each day, the Sun moves about 1 degree eastward along the celestial sphere, following a path called the **ecliptic**. The constellations along the ecliptic make up the **zodiac**.

In astrology, the ecliptic is divided evenly among twelve constellations: Gemini, Cancer, Leo, Virgo, Libra, and so on. The dates for each sign actually correspond to the positions of the constellations a very long time ago. Since then, the **precession of the equinoxes**, which is explored in another exercise, has changed the dates corresponding to each astrological sign. Furthermore, astronomers have redefined the boundaries of each constellation in a way that does not correspond exactly to the traditional boundaries. As a result, the astrological dates for the Sun's position along the zodiac do not correspond to the dates in modern astronomy.

SETUP

- Start Starry Night.
- Stop the flow of time with the STOP TIME button (■).
- Set up the following date: 12:00:00 P.M. July 1, 2007.
- Open the FIND menu item by clicking on the tab to the left.
- Click the ☑ box for the Sun. Make sure the other Solar System objects are not selected.
- Click on the small **V** button to the left of the word SUN to show a dropdown menu. Click CENTER; this will aim your point of view so that the Sun is centered along the line of sight.
- Click the FIND tab again to close.
- Click the OPTIONS tab to open.
  - Under GUIDES, have all options off (unclicked).
  - Under LOCAL VIEW, turn off DAYLIGHT and turn on LOCAL HORIZON. All others should be off.
  - Under SOLAR SYSTEM, turn PLANETS-MOONS on. All other options should be off.
  - Under STARS, turn STARS on and MILKY WAY off.
  - Under CONSTELLATIONS, turn BOUNDARIES, LABELS, and STICK FIGURES on. All other options should be off.
- Click the OPTIONS tab again to close it.

- Set the TIME FLOW RATE to be one day. The units ("days") are selected with the dropdown menu, and you can set "1" by clicking on it.
- At the top right, use the + and − keys to zoom out as necessary. At this point you should see that the Sun is located within the boundaries of Gemini.
- Now click the STEP TIME BACKWARD button (◀|) once. The TIME AND DATE display at the upper left should read noon on the previous day (June 30, 2007). Keep clicking backward until the Sun is well within Taurus. Experiment with going backward and forward one day at a time using this button and the STEP TIME FORWARD (|▶) button until the Sun is on the boundary between Taurus and Gemini.

ACTIVITY 1 – THE SUN AND THE ZODIAC

1. Use the STEP TIME FORWARD button to advance the time, and note when the Sun enters a particular constellation. Keep track of the number of days that the Sun spends in each constellation. You will be filling out a table that begins like this:

Constellation	Date Sun enters	Days spent
Gemini	6/22/2007	29
Cancer	7/21/2007	21
. . . etc.		

After a year, you will see that the Sun enters Gemini again.

ACTIVITY 2 – ASTROLOGICAL DATES

2. Compare your table of dates to the traditional astrological dates in the table that follows. What differences do you find?

Constellation	Date Sun enters
Gemini	May 21
Cancer	June 21
Leo	July 21
Virgo	August 21
Libra	September 21
Scorpius	October 21
Sagittarius	November 21
Capricornus	December 21
Aquarius	January 21
Pisces	February 21
Aires	March 21
Taurus	April 21

ACTIVITY 3 – DATES IN THE DISTANT PAST

Reset the date to May 21, 2007. Now enter A.D. 1500 for the year. You will notice that the orientation of the constellations has changed as a consequence of precession.

Using the STEP TIME FORWARD (|▶) button, you will see that, in this year, the Sun entered Gemini about June 5.

Reset the date to even earlier dates and use the time step key to find a year when the Sun entered Gemini about May 21. (Hint: You may have to go back to B.C. for this.) There's no need to be exact – use century years (e.g., A.D. 100, A.D. 1, 100 B.C.) until you find the right date.

3. In what year did the Sun enter Gemini about May 21?
4. Why does the Sun only pass through the constellations of the Zodiac? For example, why doesn't it ever pass through Ursa Major?

Name: _____

Class/Section: _____

Starry Night Student Exercise – Answer Sheet
Motion of the Sun along the Ecliptic

1. Table of dates:

Constellation	Date Sun enters	Days spent
Gemini		
Cancer		

2. What differences do you find between the traditional and the modern astronomical dates?

_____

_____

_____

3. In what year did the Sun enter Gemini about May 21?

_____

4. Why does the Sun only appear to pass through the constellations of the Zodiac? For example, why doesn't it ever appear to pass through Ursa Major?

_____

_____

Motion of the Moon

GOALS

- To investigate the apparent motion of the Moon during each month
- To determine the interval between successive moonrises
- To measure the Moon's orbital period and the time for lunar phases to repeat

READING

- Section 2.4 – The Motions and Phases of the Moon

In this exercise, we will use Starry Night to determine the length of the sidereal and synodic periods of the Moon. We will also determine the amount of time between moonrises on successive nights.

The Moon completes one orbit around Earth each month. The sidereal month is the time it takes for the Moon to return to a particular direction with respect to the stars on the celestial sphere. The word **sidereal** refers to the stars.

The period between successive phases of the Moon is called the Moon's **synodic** period. It is longer than the sidereal period because Earth is in orbit around the Sun. The difference between the synodic and sidereal periods is illustrated in Figure 2.24 of the text.

The lengths of the sidereal and synodic periods are quoted in the textbook. In this exercise, we will obtain the periods by actual measurement.

SETUP

- Start Starry Night.
- Stop the flow of time with the STOP TIME button (■).

- Set up the following date: 3:00:00 A.M. December 12, 2006.
- Make your horizon flat and free of obstructions (if it is not already). Using the dropdown menus, select OPTIONS / OTHER OPTIONS / LOCAL HORIZON. In this window, click the button next to FLAT and click OK.
- Aim to the eastern point of the horizon (space bar + E button).
- Click the OPTIONS tab to open.
  ◦ Under GUIDES, have all options off.
  ◦ Under LOCAL VIEW, turn off DAYLIGHT, turn on LOCAL HORIZON, and turn all others off.
  ◦ Under SOLAR SYSTEM, turn PLANETS-MOONS on, turn all other options off.
  ◦ Under STARS, turn STARS on and MILKY WAY off.
  ◦ Under CONSTELLATIONS, turn LABELS and STICK FIGURES on, all other options off.
- Open the FIND menu item by clicking on the tab to the left.
  ◦ Click the ☑ box for the Sun. Do the same for the box to the left of the word MOON in the list of objects. If the Moon option is not visible, click on the little + next to Earth.
  ◦ Make sure the other Solar System objects are not selected.
- Click the FIND tab again to close.
- Set the TIME FLOW RATE to be two minutes. Don't start the flow of time yet.

ACTIVITY 1 – TIME OF MOONRISE ON SUCCESSIVE NIGHTS

- Depending on your viewing location, you should see that the Moon is located near the horizon. It may be above the horizon, which means it rose earlier than 3:00 A.M. local time. If it is below the horizon, it will rise later.

- If the Moon is above the horizon, use the STEP TIME BACKWARD button (◀|). Keep going until the Moon is located right on the horizon. This will give you the time of moonrise. Note this time.
- Alternatively, if the Moon has not yet risen, use the STEP TIME FORWARD (|▶) button and record the time of moonrise.
- Advance the time using the RUN TIME FORWARD button (▶). Let one day (24 hours) pass; then stop the flow of time with the STOP TIME button (■). Then step the time forward or backward, as appropriate, until the Moon is on the eastern horizon. Record the time of moonrise.
- Do this for five days (i.e., until moonrise on December 17). You should note that the Moon rises later each night.

1. Fill out a table showing the time of moonrise on each night. It will look like this:

Night	Date	Time of moonrise	T
0	Dec. 12	1:57 A.M.	. . .
1	Dec. 13	2:54 A.M.	24 hours, 53 minutes
2	. . . etc.		
3			
4			
5			
		Average value of T	

In the last column of the table, fill in the interval between moonrises. The exact times you get for the moonrise will be different, as they depend on your viewing location.

2. What is the average amount of time that the Moon rises later each night? (Compute the average value after subtracting 24 hours from the values in the right column of your table.)

ACTIVITY 2 – THE MOON'S SIDEREAL PERIOD

- Reset the date to December 12, 2006, and the time to 3:00 A.M.
- Click on the S button toward the upper right to change your view of the southern horizon. (Students in the Southern Hemisphere should click the N button.)
- Using the time buttons, adjust the time so that the Moon is above the southern horizon, about halfway between the bottom and the top of the window.
- Point at a star near the Moon with the cursor, and note that the cursor changes to the heads-up display shape (▤).

- While holding down the shift key, click on the star. The name of the star will appear.
- If the label "The Moon" disappears, open the FIND tab and click on the box to the left of the word Moon to make it come back.
- Close the FIND tab by clicking on the word FIND.
- Write down the date and the time.
- Set the TIME FLOW RATE to be one sidereal day. Click once on STEP TIME FORWARD (|▶). You will note that the Moon has moved eastward against the stars. Keep clicking until the Moon returns closest to the star you marked. Write down the date and the time.

3. Using your measurements, how long does it take for the Moon to return to the same position compared to the stars? (If the closest approach is midway between two dates, estimate the result.)
4. What is the value of the sidereal period as stated in the textbook? How accurate was your answer in the previous problem?

ACTIVITY 3 – THE MOON'S SYNODIC PERIOD

- Select the OPTIONS tab.
  ◦ Under GUIDES, ALT-AZ GUIDES, select MERIDIAN, other options off.
  ◦ Under LOCAL VIEW, turn on DAYLIGHT.
- Click the FIND tab, and click the checkbox for the Sun. Make sure that the checkbox for the Moon is still checked.
- Close the tabs by clicking on FIND again.
- Set TIME FLOW RATE to two minutes.
- Advance the time until the Sun is located at the local meridian.
- Set the TIME FLOW RATE to one day.
- Click STEP TIME FORWARD (|▶) or STEP TIME BACKWARD (◀|) until the Moon is near the Sun. Write down this date. At this time, there is a New Moon.
- Click forward one day. Note change of position between the Moon and the Sun. Keep clicking until the Moon is closest to the Sun again, about a month later. Write down this date.

5. Using your measurements, what is the interval between times that the Moon is closest to the Sun on the sky?
6. What is the value of the Moon's synodic period in the text? How accurate was your answer in the previous problem?
7. What is the cause of the difference between the Moon's sidereal period and its synodic period? Why is one longer than the other?

Name: _____

Class/Section: _____

Starry Night Student Exercise – Answer Sheet
Motion of the Moon

1. Table of moonrise times:

Night	Date	Time of moonrise	T
		Average value of T	

2. What is the average amount of time that the Moon rises later each night?

_____

3. How long does it take for the Moon to return to the same position compared to the stars?

_____

4. What is the value of the sidereal period as stated in the textbook? How accurate was your answer in the previous problem?

_____

5. How long does it take for the Moon to return to the same position compared to the Sun?

_____

6. What is the value of the Moon's synodic period in the text? How accurate was your answer in the previous problem?

_____

7. What is the cause of the difference between the Moon's sidereal period and its synodic period? Why is one longer than the other?

_____

_____

Precession

GOAL

- To demonstrate the precession of Earth's axis

READING

- Section 2.3, last section – Earth's Axis Wobbles, and the Seasons Shift through the Year

We look more closely at the location of the north celestial pole (NCP) with respect to the stars, and see that it slowly changes over many years. This is due to the **precession** of Earth's axis. In the text, the precession is described in terms of the changing location of the equinoxes, but in this exercise we will focus on the pole. The key concepts for precession are illustrated in Figure 2.20.

Recall that the NCP is not an object in space, but rather a direction. An observer at Earth's North Pole will see that the NCP is at the zenith (i.e., straight up). An observer at another latitude will see the celestial pole at an angle above the northern horizon (e.g., Figure 2.8).

Earth's rotation axis wobbles slowly. Over a period of about 26,000 years, it traces out a circle on the sky that is about 47 degrees in diameter. In this exercise, we watch the changing position of the NCP over a few thousand years.

SETUP

- Start Starry Night.
- Stop the flow of time with the STOP TIME button (■).
- Using the dropdown menus, select OPTIONS / VIEWING LOCATION. A box labeled VIEWING LOCATION will appear.
  - Select the LATITUDE / LONGITUDE tab.
  - In the upper dialog box (LATITUDE), remove the current coordinate and type **55 N**.
  - Click the GO TO LOCATION button at the lower right of the dialog box.
  - Your viewing location will now shift. You can speed up the process by hitting the space bar while the motion is taking place. At this point, you may see the sky or you may be looking at the ground.
  - Click on the N button toward the upper right to point your gaze toward the north.
- Click the OPTIONS tab to open.
  - Under CONSTELLATIONS, turn on BOUNDARIES, LABELS, and STICK FIGURES; all others should be off.
  - Under GUIDES, have all CELESTIAL POLES on and all other options off.
  - Under LOCAL VIEW, turn on LOCAL HORIZON; all others should be off.
  - Under SOLAR SYSTEM, all options should be off.
  - Under STARS, turn STARS on and MILKY WAY off.
- Close the OPTIONS tab by clicking on the word OPTIONS.

ACTIVITY 1 – THE DISTANCE BETWEEN POLARIS AND THE NORTH CELESTIAL POLE

- Identify Polaris and click on it with the left mouse button. (Polaris is the bright star at the end of the handle of the Little Dipper in the constellation Ursa Minor; use the stick figures for the constellation as a guide.) Do NOT use the CENTER option (found on the right-click dropdown menu) to center on Polaris.
- Zoom in toward the NCP using the + or − buttons (top right) or the cursor wheel. You may have to use

the handgrab tool to move the NCP so that it is near the center of the window. Zoom in so that the NCP and Polaris are well separated.

- Use the angle measuring tool to find the separation between the NCP and Polaris. To do this, move the cursor over Polaris until the cursor changes to the heads-up display shape (▤). While the cursor has this shape, click on Polaris and drag the mouse toward the NCP. When the tip of the cursor is located at the little cross representing the NCP, you will be able to read off the separation. The angular separation is listed in degrees, minutes, and seconds. (One minute is 1/60 degree; one second is 1/60 of a minute or 1/3,600 of a degree.)

1. What value do you get for the current separation between Polaris and the NCP?

ACTIVITY 2 – WHEN POLARIS IS CLOSEST TO THE NCP

- Change the time flow rate to 366 sidereal days.
- Click RUN TIME FORWARD (▶). You will see that the position of the NCP relative to Polaris slowly changes.
- When you estimate that Polaris is closest to the NCP, click STOP TIME (■). You may wish to go backward and forward until you decide that the two are closest together.

2. Approximately when (within 50 years) will Polaris and the NCP be closest together?

3. What will the separation be compared to now, in percent? (Take a ratio of the separation values.)

ACTIVITY 3 – THE SEPARATION IN THE PAST

- Long before compasses were invented, navigators in the Northern Hemisphere used Polaris as the "North Star" to orient themselves. Set the Time and Date values so that they show some date in the year A.D. 1.

4. What was the separation between Polaris and the NCP in the year A.D. 1?

- Consider the stars connected by the stick figures in the constellations of Ursa Minor and Draco. Try to identify stars in these two constellations that were closer to the NCP than Polaris was, and so would make a better North Star. Select a candidate star by moving the cursor over it until the cursor changes to the heads-up display shape (▤). Left-click on the star to reveal its name. Then use the angle separation tool to find the separation between this star and the NCP.

5. List two or three stars that were closer than Polaris to the NCP in the year A.D. 1. What was the angular separation between each star and the NCP?
6. Does the change in the position of the north celestial pole have any effect on the other stars in the sky besides Polaris? Why or why not?

Name: _____

Class/Section: _____

Starry Night Student Exercise – Answer Sheet
Precession

1. What value do you get for the current separation between Polaris and the NCP?

   _____

2. Approximately when (within 50 years) will Polaris and the NCP be closest together?

   _____

3. What will the separation be compared to now, in percent? (Take a ratio of the separation values.)

   _____

4. What was the separation between Polaris and the NCP in the year A.D. 1?

   _____

5. List two or three stars that were closer than Polaris to the NCP in the year A.D. 1. What was the angular separation between each star and the NCP?

   _____

   _____

   _____

6. Does the change in the position of the north celestial pole have any effect on the other stars in the sky besides Polaris? Why or why not?

   _____

   _____

Kepler's Laws

GOAL

- To demonstrate Kepler's second and third laws for planetary orbits

READING

- Section 3.3 – An Empirical Beginning: Kepler's Laws

Based on accurate data from Tycho Brahe, Kepler advanced three empirical **laws** describing the orbits of the planets. In particular, he came up with rules relating the size of an orbit to the orbital period (the third law), and described the speed a body follows along its orbital path (the second law). The exact wording of these laws can be found in section 3.2 of the textbook. The word **empirical** means that the laws are derived from observations and not from physical theory.

Each planet orbits the Sun along a path that is in the shape of an **ellipse**. The size of an orbit is given by the length of the **semimajor axis**. This is half the longest dimension of the orbit. Each planet will have a minimum and a maximum distance from the Sun called, respectively, the **perihelion** and the **aphelion**.

Most of the major planets have orbits that are almost circular; they travel at nearly a constant speed along the orbit. In this exercise, we will create a new asteroid and give it a very elongated orbit, which will make its changing speed more apparent.

SETUP

- Start Starry Night.
- Stop the flow of time with the STOP TIME button (■).

- Click the OPTIONS tab to open.
  - Under GUIDES, turn all options off.
  - Under LOCAL VIEW, turn all options off.
  - Under SOLAR SYSTEM, select PLANETS, select ASTEROIDS, and turn all other options off.
  - Under STARS, turn all options off.
  - Under CONSTELLATIONS, turn all options off.
  - Under DEEP SPACE, turn all options off.
- Click on the FIND tab to open.
  - Double-click on the word SUN. This will shift your gaze so you are pointed at the Sun and will put a label near the Sun.
  - Click the box to the left and the first box to the right of Earth. This will show Earth's orbit and label its position. Leave the FIND tab open for now.
- Select OPTIONS / VIEWING LOCATION from the dropdown menus at the top of the Starry Night window. Click on the down-arrow (▼) next to the display box for VIEW FROM. Select STATIONARY LOCATION.
- Under CARTESIAN COORDINATES, replace the current numbers with
  - X: 0 au
  - Y: 0 au
  - Z: 4 au

then hit the GO TO LOCATION button. The view will now shift so that you are looking down on the Sun and Earth's orbit from a position 4 au above the Sun.

ACTIVITY 1 – KEPLER'S THIRD LAW

- Below the list of the planets, find the word ASTEROIDS. If there is a + next to the word, click on the plus to display the list of asteroids.

- We will create a new asteroid that has an elongated orbit. Using the dropdown menus at the top of the Starry Night window, select FILE / NEW ASTEROID ORBITING SUN. A subwindow labeled ASTEROID: UNTITLED will appear.
  - There will be a dialog box at the top that has the word UNTITLED in it. Replace what is there with a capital **X**. This is the name of the new asteroid. You can call it something else, as long as it is not the same as one of the asteroids already on the list.
  - The subwindow will show a series of slider bars under a tab called ORBITAL ELEMENTS. To the right of each slider bar is a dialog box in which you can enter numbers manually. Enter the following numbers:
    - Mean distance (a): 1.0
    - Eccentricity (e): 0.75
  - Leave the other numbers alone. Close the subwindow by clicking on the × at the top right corner. You will be asked whether you want to save the new information. Select SAVE.
- Immediately above the ASTEROIDS list is the USER CREATED list. Expand this and find the name of the asteroid you created (X). Click the boxes to the left and right of the name to show the orbit and the position of asteroid X.

1. The value of a you entered is the length of the semimajor axis of X's orbit. This asteroid has the same value of a as does the Earth. Use Kepler's third law, which can be found in section 3.3 of the textbook, to determine its orbital period. What is the orbital period of asteroid X?
2. Should the period of the orbit depend on the orbital eccentricity e?
3. If we had created an orbit with $a = 4$ au, what would the period be?

ACTIVITY 2 – THE PERIOD OF X'S ORBIT

- Unclick the left button next to Earth in the FIND tab. This will remove the label for Earth from the display. You can remember which orbit is Earth's by recalling that Earth's is the round one.
- Click on the word FIND to close the FIND tab.
- Using the zoom in/out bottoms (+ or −) or the mousewheel, zoom in until Earth's orbit fills the Starry Night display. You will notice that there are two places where X's orbit crosses Earth's.
- Set the TIME FLOW RATE to one day. Using the motion buttons, advance or reverse the time until X crosses Earth's orbit. Stop the flow of time.

- Use the STEP TIME FORWARD (▶|) or STEP TIME BACKWARD (◀|) buttons until you find the exact date on which X was closest to Earth's orbit.
- Write down the date from the TIME AND DATE display.
- Now hit RUN TIME FORWARD (▶) and let X orbit the Sun completely. Stop the time when it has gone all the way around its orbit and is crossing Earth's orbit again. You have to pick the original crossing point and ignore the other one. Line up X with Earth's orbit as you did in the previous step. Write down the date.

4. How long did it take for X to complete an orbit? Does your answer agree with your computation in question 1?
5. Is X moving faster at perihelion or at aphelion?
6. State Kepler's second law. Does your answer for question 5 agree with this law?

ACTIVITY 3 – INSIDE AND OUTSIDE

- Now determine the date on which X crosses the Earth's orbit going inward and the date when it crosses Earth's orbit going outward.

7. How many months does X spend inside Earth's orbit? How many months does it spend outside Earth's orbit?

ACTIVITY 4 – KEPLER'S SECOND LAW

- Click on the FIND tab to open.
- Point your cursor at the name X under the USER CREATED column. Right-click on X, and select SHOW INFO.
- Under POSITION IN SPACE, find the display for DISTANCE FROM THE SUN. Use the time flow buttons to move X until it is at perihelion. Write down the distance. Do the same for aphelion.

8. What is the distance from the Sun at perihelion and at aphelion?
9. If the semimajor axis for X is 1 au, what is the length of the major axis? Does this agree with the numbers you measured in question 8?
10. If you wanted to find all of the potential hazardous asteroids, like Asteroid X, which have orbits that cross Earth's orbit, where in the Solar System are you most likely to find them at any given time? Thus, how would you focus your observational search for such objects? (Hint: Think of what Kepler's second law says about elliptical orbits.)

Flying to Mars

GOAL

- To create an orbit that takes a spacecraft from Earth to Mars

READING

- Section 3.3 – An Empirical Beginning: Kepler's Laws
- Section 6.5 – Getting Up Close with Planetary Spacecraft

Kepler's laws of planetary motion have many applications besides explaining the orbits of the planets, asteroids, and comets. In this exercise, we will see how we can set up an orbit that will carry a spacecraft from Earth to the other planets. We will do this by creating an asteroid in the Starry Night database and adjusting the parameters of its orbit until it connects Earth's orbit with that of Mars. Such an orbit is called a Hohmann transfer orbit, and generally it minimizes the amount of fuel needed to reach another planet. We will also find that spacecraft have to be launched at particular times in order to reach their destination.

We will also consider an important complication. If the spacecraft has a human crew aboard, the timing gets very tricky indeed! Not only does the spacecraft have to be launched at just the right time to reach Mars, but Earth might not be in the right place when the crew wants to return.

SETUP

- Start Starry Night.
- Stop the flow of time with the STOP TIME button (■).

- Click on the OPTIONS tab to open.
  - Under GUIDES, turn all options off.
  - Under LOCAL VIEW, turn all options off.
  - Under SOLAR SYSTEM, select PLANETS and ASTEROIDS and turn all other options off.
  - Under STARS, turn all options off.
  - Under CONSTELLATIONS, turn all options off.
  - Under CONSTELLATIONS, turn all options off.
  - Under DEEP SPACE, turn all options off.
- Click on the FIND tab to open.
  - Double-click on the word SUN. This will shift your gaze so that you are pointed at the Sun and will put a label near the Sun.
  - Click the box to the left and the first box to the right of Earth. This will show Earth's orbit and label its position. Do the same for Mars. Leave the FIND tab open for now.
- Select OPTIONS / VIEWING LOCATION from the dropdown menus at the top of the Starry Night window. Click on the down-arrow (▼) next to the display box for VIEW FROM. Select STATIONARY LOCATION.
- Under CARTESIAN COORDINATES, replace the current numbers with
  - X: 0 au
  - Y: 0 au
  - Z: 4 au
  Then hit the GO TO LOCATION button. The view will now shift so that you are looking down on the Sun and Earth's orbit from a position 4 au above the Sun.
- If you completed the exercise "Kepler's Laws," you will have created an asteroid called X. If not, then do the following steps:
  - Using the dropdown menus at the top of the Starry Night window, select FILE / NEW ASTEROID

ORBITING SUN. A subwindow labeled ASTEROID: UNTITLED will appear.

There will be a dialog box at the top that has the word UNTITLED in it. Replace what is there with a capital **X**.

The display should show a series of slider bars under a tab called ORBITAL ELEMENTS. To the right of each slider bar is a dialog box in which you can enter numbers manually. Enter the following numbers:

> Mean distance (*a*): 1.00
> Eccentricity (*e*): 0.75
> Epoch: 2451545.0 (Julian day)

Leave the other numbers alone. Close the subwindow by clicking on the × at the top right corner. You will be asked whether you want to save the new information. Select SAVE.

- Now find the entry under the USER CREATED list called the X. Click the boxes to the left and to the right of the name X. This will show the orbit and position of X.

ACTIVITY 1 – THE TRANSFER ORBIT

- Set the date in the TIME AND DATE window to read June 28, 2003.
- Using the zoom in / out buttons or your mousewheel, zoom in until the orbit of Mars fills the display window.
- Under the FIND tab, turn off the names for Earth and Mars by clicking the boxes to the left of the names of these planets. Leave the boxes to the right of each name checked. You will still see the location of each planet and its orbit.
- Find the name X under the list of Asteroids. Right-click on the name and select EDIT ORBITAL ELEMENTS.
- Using the various slider bars, adjust the MEAN DISTANCE (*a*), the ECCENTRICITY (*e*), and the ARGUMENT] OF PERICENTER (*w*) values until the orbit of X touches Earth's orbit at X's perihelion and Mars' orbit at X's aphelion. In particular, you want the perihelion of X to be at the same place as Earth was on June 28, 2003. First start with values close to
  ○ Mean distance (*a*): 1.25 au
  ○ Eccentricity (*e*): 0.20
  ○ Argument] of pericenter (*w*): 270°

Using the slider bars, make small adjustments so that the orbit of X connects Earth's orbit with Mars' orbit.
- Now adjust the value of the MEAN ANOMALY (*L*) until the asteroid X is located where Earth was on June 28, 2003. It should be near 22°.
- Write down the values for *a*, *e*, *w*, and *L*.
- When the orbit of X runs from the Earth to Mars, click on the × at the upper right corner of the subwindow to close it.

1. What value did you get for the eccentricity (*e*) for X's orbit?
2. What is the mean distance (*a*)?
3. What is the value of the argument of pericenter (*w*)?
4. What is the value of the mean anomaly (*L*)?

ACTIVITY 2 – TIME TO FLY TO MARS

- Set the TIME FLOW RATE to one day. Start the flow of time with ▶ and let it run. You will see that X follows the orbit you set up previously. You should see that the spacecraft gets close to Mars after several months. Don't worry if it misses a little bit – it's not easy to set up an exact orbit that places the spacecraft exactly at Mars' location (NASA has fancy computers for that!). Use the STOP TIME (■) and the STEP TIME (▐▶ or ◀▐) buttons to settle on a time when X is closest to Mars.

5. Approximately when does the spacecraft arrive in the vicinity of Mars?

ACTIVITY 3 – COMING HOME

- Now continue the flow of time with the ▶ button. The spacecraft will return to Earth's orbit.

6. When does X return to Earth's orbit? Is Earth there at the time?
7. What is the orbital period of X?
8. Suppose X was a spacecraft with a crew on board. As a mission planner, what might you have to do to make sure the crew gets back to Earth safely?

Name: _____

Class/Section: _____

Starry Night Student Exercise – Answer Sheet
Flying to Mars

1. What value did you get for the eccentricity (e) for X's orbit?

_____

2. What is the mean distance (a)?

_____

3. What is the value of the argument of pericenter (w)?

_____

4. What is the value of the mean anomaly (L)?

_____

5. Approximately when does the spacecraft arrive in the vicinity of Mars?

_____

6. When does X return to Earth's orbit? Is Earth there at the time?

_____

7. What is the orbital period of X?

_____

8. Suppose X was a spacecraft with a crew on board. As a mission planner, what might you have to do to make sure the crew gets back to Earth safely?

_____

_____

_____

The Moons of Jupiter

GOAL

- To investigate historical observations of Jupiter and its four large moons, and to use these observations to deduce the motion of these moons about the planet

READING

- Section 3.2 – An Empirical Beginning: Kepler's Laws
- Section 11.1 – Moons in the Solar System

In the year 1610 Galileo first turned a telescope to the planet Jupiter. What he found there were four small "stars" that appeared to move along with Jupiter, constantly changing their position relative to the planet. Over the course of several weeks in January of that year he methodically recorded the changing positions of these "stars," sometimes observing them more than once per night. That he was able to accurately record the positions of what we now understand to be four moons is especially amazing considering the rudimentary telescope he used.

By using his observations Galileo was able to deduce that these moons were in orbit around Jupiter and were not just stars in the background. This was a revolutionary idea at a time when everything was thought to orbit the Earth. But this was not a simple deduction to make. In this lab we will first recreate Galileo's observations using the dates and times he recorded. Then we will find the moons' orbital periods and deduce the relation between their distances, speeds, and periods.

SETUP

- Start Starry Night.
- Stop the flow of time with the STOP TIME button (■).
- Click on the OPTIONS tab to open.
  - Under GUIDES, all options should be off.
  - Under LOCAL VIEW, turn on DAYLIGHT and LOCAL HORIZON. All others should be off.
  - Under SOLAR SYSTEM, turn on PLANETS-MOONS. All others should be turned off, including PLANETS-MOONS LABELS.
  - Under STARS, turn on STARS, turn off MILKY WAY.
  - Expand the STARS options by clicking the heads-up display shape ▤.
  - Move the cursor over LIMIT BY MAGNITUDE. An option box labeled LIMIT BY MAGNITUDE OPTIONS will become visible. Click on this.
  - A window titled STARS OPTIONS will appear. In this window click the ☑ box next to LIMIT STARS BY MAGNITUDE.
  - In the box to the right of the slider bar, type in **7.00**.
  - Click OK to exit this window.
  - Under DEEP SPACE, turn all options off.
- Click the OPTIONS tab again to close.

ACTIVITY 1 – GALILEO'S OBSERVATIONS

- Using the dropdown menus, select OPTIONS / VIEWING LOCATION. A box labeled VIEWING LOCATION will appear.
  - Click the button next to SHOW WHERE.
  - In the text box to the right, type "**Padova**," which will bring up the location name Padova, Veneto, Italy, the city where Galileo made his observations.

- Click on this location name to select it.
  - Click the Go to Location button in the lower right corner of the dialog box.
- Click the Find tab to open.
  - Click the small **V** button to the left of the checkbox for Jupiter to show a dropdown menu. Click Center; this will aim your point of view so that Jupiter is in the center of your screen.
- Click the Find tab again to close.
- Set the following date and time: 8:00:00 P.M. January 7, 1610 A.D.
- At the top right, use the + button to zoom in until the display reads 1° × 1°.

1. Your view is now set up to display what Galileo would have seen through his telescope on January 7, 1610. In fact, his view was much more limited due to the small field of view of this telescope. His telescope also did not have a large magnification, which is why Jupiter looks so small, making it a challenge to identify the moons. His first observation as taken from his own records is completed for you in the table. Notice that he only recorded three of the moons. Continue filling in the table using the dates and times of Galileo's next six observations. (Note that January 09 was skipped due to poor weather.)

Date	Time	Sketch
1/07/1610	8:00 P.M.	＊ ＊ ○ ＊
1/08/1610	8:00 P.M.	
1/10/1610	8:00 P.M.	
1/11/1610	8:00 P.M.	
1/12/1610	8:00 P.M.	
1/13/1610	8:00 P.M.	
1/14/1610	8:00 P.M.	

2. In his early observations Galileo only saw three moons. It was not until January 13, 1610, that he recorded the fourth. Looking at your observations, can you propose an explanation for why Galileo may not have immediately noticed all four moons? Use one of your own observations to support your answer.

ACTIVITY 2 – THE ORBITS OF JUPITER'S MOONS

- Using the dropdown menus, select Options / Viewing Location. A box labeled Viewing Location will appear.
  - Click on the down-arrow (▼) next to the display box for View from. Select Stationary Location.
- Under Cartesian Coordinates, replace the current numbers with
    X: 1 au
    Y: 0 au
    Z: 0 au
  Then click the Go to Location button. The view will now shift so that you are looking out into space at a position 1 au from the Sun.

- Click the Find tab to open.
- Click the small **V** button to the left of the checkbox for Jupiter to show a dropdown menu. Click Center; this will aim your point of view so that Jupiter is in the center of your screen.
- Click the ⊞ box next to Jupiter. This will display the four Galilean moons (Io, Europa, Ganymede, and Callisto).
- Click the checkbox to the left of each of these moons to label them on the display.
- Click the Find tab again to close.
- At the top right, use the + button to zoom in until Jupiter and all four of the Galilean moons can be seen distinctly.
- Set Time Flow Rate to 1 hour. Press the Run Time Forward button (▶) and let it run. You will see the four labeled moons move back and forth around Jupiter. Press the Stop Time button (■) once you have watched several orbits of the moons.

3. Galileo was able to deduce that these four moons moved in nearly circular orbits around Jupiter without the benefit of this time-lapse view. Seeing the moons in motion, what evidence can you provide that tells you they do move in a circular motion, even though you cannot see the circle itself?

4. Which moon stays the closest to Jupiter? Which moon moves the farthest from Jupiter?

5. Which moon circles Jupiter with the fastest speed? Which moves the slowest?

- Click the Step Time Forward button (|▶) until the moon Io is as far to the left of Jupiter as it can get. In this position Io is at its greatest elongation from Jupiter. Record the date and time in the table under Elongation 1. Because you are observing from space, the times will be shown as a 24-hour clock.
- Continue clicking the Step Time Forward button (|▶) until Io returns to the same position. Record the date and time under Elongation 2.
- Repeat these two instructions for Europa, Ganymede, and Callisto.
- Find the amount of time between these two sequential elongations for each moon. This will be the orbital period of the moon.

6. Fill out the table, showing sequential elongations for each moon. It will look like this:

Moon	Elongation 1	Elongation 2	Period
Io	6:00, Jan 27	0:00, Jan 29	1 day, 18 hours
Europa			
Ganymede			
Callisto			

7. Using what you have observed of the moons' motions, state the relation between a moon's distance from Jupiter, its orbital speed, and its orbital period.

Name: _____

Class/Section: _____

Starry Night Student Exercise – Answer Sheet (continues on back)
The Moons of Jupiter

1. Table of observations of Jupiter's moons:

Date	Time	Sketch
1/07/1610	8:00 P.M.	* * ○ *
1/08/1610	8:00 P.M.	
1/10/1610	8:00 P.M.	
1/11/1610	8:00 P.M.	
1/12/1610	8:00 P.M.	
1/13/1610	8:00 P.M.	
1/14/1610	8:00 P.M.	

2. Looking at your observations, can you propose an explanation for why Galileo may not have immediately noticed all four moons? Use one of your own observations to support your answer.

_____

_____

_____

3. Seeing the moons in motion, what evidence can you provide that tells you they do move in a circular motion, even though you cannot see the circle itself?

_____

_____

4. Which moon stays the closest to Jupiter? Which moon moves the farthest from Jupiter?

_____

5. Which moon circles Jupiter with the fastest speed? Which moves the slowest?

_____

6. Table of elongations and periods:

Moon	Elongation 1	Elongation 2	Period
Io	6:00, Jan 27	0:00, Jan 29	1 day, 18 hours
Europa			
Ganymede			
Callisto			

7. Using what you have observed of the moons' motions, state the relation between a moon's distance from Jupiter, its orbital speed, and its orbital period.

_____

_____

_____

STUDENT EXERCISE	The Rings of Saturn

GOAL

- To investigate the changing appearance of Saturn's rings and how this allows us to make new discoveries in the Saturnian system

READING

- Section 11.3 – The Discovery of Rings around the Giant Planets

The four large Jovian planets in our Solar System share a common set of characteristics, such as large gaseous atmospheres, an abundance of moons, and ring systems.

Saturn's ring system is the largest and only one visible through Earth-bound telescopes. From our point of view they appear to be large disks orbiting above the planet's equator. Saturn, like Earth, is tilted relative to the plane of the Solar System. This means that we will see the rings in different inclination angles at different points of the year, depending on the tilt of the planet and Earth's position in our orbit. Approximately twice every Saturn orbit around the Sun, Earth will pass through the plane of Saturn's rings. On these days, from Earth's point of view the rings will have a tilt (inclination) of zero degrees, so that we will see them edge-on. These **ring plane crossings** are times when astronomers can learn more about the thickness and physical arrangement of the rings. Also, because of the lack of glare from the highly reflective rings, these are ideal times to search for previously unknown moons of Saturn. In this exercise we will investigate the changing appearance of Saturn's rings as seen from Earth.

SETUP

- Start Starry Night.
- Stop the flow of time with the STOP TIME button (■).
- Set the date in the TIME AND DATE window to read January 1, 2009.
- Click the OPTIONS tab to open.
  - Under LOCAL VIEW, turn off all options.
  - Under SOLAR SYSTEM, turn on PLANETS-MOONS. All other options should be off.
- Click the OPTIONS tab again to close it.
- Click the FIND tab to open.
  - Click the ☑ box for Saturn.
  - Click on the small **V** button to the left of the checkbox for Saturn to show a dropdown menu. Click CENTER; this will aim your point of view so that Saturn is in the center of your screen.
- Click the FIND tab again to close.
- Set the TIME FLOW RATE to be one day. The units ("days") are selected with the dropdown menu, and you can set "1" by clicking on it. (Make sure this is NOT "sidereal d.") Don't start the flow of time yet.
- At the top right, in the ZOOM box, click the + key to zoom in until the display reads 1′ × 1′.

ACTIVITY 1 – FINDING SATURN'S RING PLANE CROSSING

1. Observe the inclination of Saturn's rings. Describe their inclination and speculate on how they might affect observations of other nearby objects.

- Press the START TIME button (▶) to run time forward and observe the changing tilt of Saturn and its rings.
- You will notice that at one point during 2009 the rings will appear edge-on and nearly vanish. Use the STOP TIME button (■) when you get close or pass this alignment, then use the STEP TIME FORWARD / BACKWARD buttons (▶| / |◀) as needed to find the exact date.

2. What is the date that Saturn's rings appear edge-on? With the rings in this orientation, how might observations of other nearby objects be affected?
3. What can you say about the thickness of Saturn's rings compared to the planet itself? (Try zooming in using the + or – buttons in the ZOOM window in the upper right to get a closer view.)

ACTIVITY 2 – SATURN'S RINGS AT MAXIMUM INCLINATION

- In the ZOOM window in the upper right, reset your zoom to 1′ × 1′.
- Set the TIME FLOW RATE to five days.
- Press the START TIME button (▶) to run time forward and observe how our view of Saturn and its rings will change over the next few years.
- When the rings appear to be tilted their maximum amount from edge-on, press the STOP TIME button (■). Use the STEP TIME FORWARD / BACKWARD buttons (▶| / |◀) until you are at the approximate date when the rings are tilted the most.

4. As you watched Saturn, its size appeared to fluctuate from slightly larger to slightly smaller. What could be causing the apparent change in size of Saturn? (Think about the orbits of Earth and Saturn and the fact that this fluctuation repeats every year.)
5. What is the approximate date at which Saturn's rings are tilted their maximum amount from edge on?
6. On this date, describe the width of the rings as compared to the diameter of Saturn.

ACTIVITY 3 – FINDING THE NEXT RING PLANE CROSSING

- Press the START TIME button (▶) to continue the flow of time.
- The tilt of the rings will again get smaller until the rings are once more edge-on. Use the STOP TIME button (■) when you get close or pass this alignment, then use the STEP TIME FORWARD / BACKWARD buttons (▶| / |◀) as needed to find the exact date.

7. What is the next date when Saturn's rings again appear edge-on?
8. What is the interval between the two dates when Saturn's rings appear edge-on? What is the date you would have predicted?

Name: _____

Class/Section: _____

Starry Night Student Exercise – Answer Sheet
The Rings of Saturn

1. Observe the inclination of Saturn's rings. Describe their inclination and speculate on how they might affect observations of other nearby objects.

   _____

   _____

2. What is the date that Saturn's rings appear edge-on? With the rings in this orientation how might observations of other nearby objects be affected?

   _____

3. What can you say about the thickness of Saturn's rings compared to the planet itself? (Try zooming in using the + or – buttons in the ZOOM window in the upper right to get a closer view.)

   _____

4. As you watched Saturn, its size appeared to fluctuate from slightly larger to slightly smaller. What could be causing the apparent change in size of Saturn? (Think about the orbits of Earth and Saturn and the fact that this fluctuation repeats every year.)

   _____

5. What is the approximate date at which Saturn's rings are tilted their maximum amount from edge-on?

   _____

6. On this date describe the width of the rings as compared to the diameter of Saturn.

   _____

7. What is the next date when Saturn's rings again appear edge-on?

   _____

8. What is the interval between the two dates when Saturn's rings appear edge-on? What is the date you would have predicted?

   _____

Pluto and Kuiper Belt Objects

GOAL

- To show why Pluto's orbit is like those of many other small bodies in the outer Solar System

READING

- Section 12.2 – Dwarf Planets: Pluto and Others

In 2006 there was a lot of publicity about new guidelines from the International Astronomical Union about the use of the word **planet**; as you probably know, Pluto doesn't fit the definition. Pluto and its companion Charon are very small, icy worlds – in their composition, they are more like giant comet nuclei than any other class of Solar System objects.

Astronomers have long suspected that the outer reaches of the Solar System may contain a large number of objects like Pluto in the distant Kuiper Belt. Several of these have been discovered in the last decade or so, and are now called **Kuiper Belt objects** (KBOs). In this exercise, we will see that Pluto is kin to the KBOs in its orbit as well as its composition.

SETUP

- Start Starry Night.
- Stop the flow of time with the STOP TIME button (■).
- Hit the OPTIONS tab to open.
  - Under GUIDES, turn all options off.
  - Under LOCAL VIEW, turn all options off.
  - Under STARS, turn all options off.
  - Under SOLAR SYSTEM, turn on ASTEROIDS, turn on PLANETS-MOONS; turn all other options off.
  - Under CONSTELLATIONS, turn off all options.
  - Under DEEP SPACE, turn off all options.
- Click on the word OPTIONS to close this tab. At the moment, there may be few objects visible, as the program is now showing only those planets that are above the horizon from your viewing location.
- Change your viewing location so that you are looking down at the Sun and can see the orbits of asteroids and planets:
  - Select OPTIONS / VIEWING LOCATION from the drop-down menus at the top of the Starry Night window. The VIEWING LOCATION subwindow will appear.
  - At the top of the subwindow, click on the down-arrow to the right of the dialog box labeled VIEW FROM. Select STATIONARY LOCATION.
  - The box will change so that you can select a fixed location in space. These coordinates are centered on the Sun. Replace the numbers in the boxes labeled CARTESIAN COORDINATES, and enter the following values:
    - X: 0 au
    - Y: 0 au
    - Z: 75 au
- Click on the GO TO LOCATION button. The view will now shift to a location above the Sun. You can hit the space bar during the motion to speed up the process.
- Click on the FIND tab to open it.
  - Double-click on the word SUN in the NAME column. This will shift the angle of your gaze so that it is centered on the Sun.
- Click on the ⊞ icon next to DWARF PLANETS.

- Turn on the names for Neptune and Pluto by clicking on the empty ☑ boxes to the left of their names.
- Turn on the orbits of Neptune and Pluto by clicking the empty ☑ boxes to the right of their names.
- Leave the FIND tab open.
- Zoom in or out using the + and – buttons at the top right of the Starry Night window. Adjust the zoom until the orbit of Pluto fills the display but is still visible all around.

ACTIVITY 1 – PERIHELION AND APHELION OF NEPTUNE AND PLUTO

Pluto can approach closer to the Sun than Neptune does, but its orbit does not cross Neptune's. Explore this by finding the minimum distance (perihelion) for Pluto, and compare it to Neptune's distance at the time Pluto was at perihelion.

- Set the TIME FLOW RATE to 20 days.
- Move your cursor to the word PLUTO in the FIND tab. Right-click on PLUTO, and select SHOW INFO. If there is a + next to POSITION IN SPACE, click on the + to display Pluto's location. Leave the FIND tab open.
- Using the TIME FLOW buttons, find the moment when Pluto was last closest to the Sun. (This was back around 1989.) Use the STEP TIME FORWARD (▶) and STEP TIME BACKWARD (◀) buttons to stop at the time when Pluto was at perihelion.

1. What was the month and year in which Pluto was at perihelion?
2. What distance was Pluto from the Sun in that month?

- Click on the FIND tab to show the names of the planets. As before, show the information for Neptune by right-clicking on the name NEPTUNE and selecting SHOW INFO.
- If there is a + next to POSITION IN SPACE, click on the + to display Neptune's position.

3. What was Neptune's distance from the Sun during that time?

ACTIVITY 2 – ORBITS IN THE OUTER SOLAR SYSTEM

- Click on the FIND tab to display the names of the planets.
- Under ASTEROIDS, click on the + button to show the names of many asteroids.
- Columns of data will appear on several asteroids. If they are not sorted by the length of the semimajor axis of their orbits, click on the word SEMIMAJOR at

the top of the data columns. If they are sorted in increasing order, there will be a little down-arrow just after the word SEMIMAJOR.

- Scroll down using the slider bar in the left subwindow. You will now display the orbits of several objects in the outer Solar System. These are members of the Kuiper Belt and are called Kuiper Belt objects (KBOs), although they are listed in Starry Night under ASTEROIDS. The box to the left of the name displays a marker for the location of each KBO. The box to the right shows the orbit for these KBOs. Click both boxes for the last four objects on the list: 2003 EL61, 2005 FY9, 2004 XR190, and 1998 WA31.
- If your version of Starry Night does not have these objects, you can enter them using the following procedure:
  ◦ Using the dropdown menus at the top of the Starry Night window, select FILE / NEW ASTEROID ORBITING SUN. A subwindow labeled ASTEROID: UNTITLED will appear.
  ◦ There will be a dialog box at the top that contains the word UNTITLED. Replace UNTITLED with **2003 EL61**.
  ◦ The display will show a series of slider bars under a tab called ORBITAL ELEMENTS. To the right of each slider bar is a dialog box in which you can enter numbers manually. Enter the following numbers:
    Mean distance (a): 43.32
    Eccentricity (e): 0.19
    Inclination (i): 28.2
  ◦ Don't worry about the values of the other parameters.
  ◦ Close the subwindow by clicking on the × at the top right corner. You will be asked whether you want to save the new information. Select SAVE.

	2005 FY9	2004 XR190	1998 WA31
Mean distance (a)	45.66	57.03	54.78
Eccentricity (e)	0.16	0.08	0.43
Inclination (i)	29.0	46.8	9.5

- Repeat the process for the other three objects using the information in the table. Then check the boxes next to the new objects in the USER CREATED list to show the orbits and positions of these outer Solar System objects.
- Close the FIND tab by clicking on the word FIND.
- These orbits are much larger than Pluto's orbit, so we need to zoom out. Using the procedure above for setting the location, move the viewpoint to
    X: 0 au
    Y: 0 au
    Z: 150 au

- Zoom in or out so that the largest orbit is just within the viewing window.
- Set the TIME FLOW RATE to 10 days. Start the flow of time. Let it run until several years of the simulation have passed. Stop the flow of time.
- Now change your point of view so that you are aligned with Neptune's orbit. Following the procedure above, change your location to
  - X: 0 au
  - Y: 150 au
  - Z: 0 au

After the view shifts, you should easily see the inclinations of the outer Solar System orbits.

4. All the other planets have small values of eccentricity and orbital inclination. Based on the values for the other objects in this exercise, is Pluto more like the KBOs or the classical planets?
5. Given the orbits of Pluto and the KBOs, if you wanted to search the sky for similar objects, would you look in the same part of the sky as you find the classical planets? Why or why not?

Name: _____

Class/Section: _____

Starry Night Student Exercise – Answer Sheet
Pluto and Kuiper Belt Objects

1. What was the month and year in which Pluto was at perihelion?

   _____

2. What distance was Pluto from the Sun in that month?

   _____

3. What was Neptune's distance from the Sun during that time?

   _____

4. Is Pluto more like the KBOs or the classical planets?

   _____

5. Given the orbits of Pluto and the KBOs, if you wanted to search the sky for similar objects, would you look in the same part of the sky as you find the classical planets? Why or why not?

   _____

   _____

Asteroids

GOALS

- To explore the orbits of some objects in the asteroid belt between Mars and Jupiter
- To examine the orbit of one asteroid that passes near Earth

READING

- Section 12.3 – Asteroids—Pieces of the Past

There are many asteroids in the **main asteroid belt** between Mars and Jupiter. In addition, some have orbits that take them near Earth. The latter are called **near-Earth asteroids**.

Figure 12.8 shows the orbits for several classes of asteroids. Asteroids have orbits that lie near the plane of the Solar System, but on average their orbits are moderately inclined to that plane. In this exercise, we will animate some of the orbits in that figure, and view them from different angles so that we can see the shapes and orientations of their orbits.

SETUP

- Start Starry Night.
- Stop the flow of time with the STOP TIME button (■).
- Hit the OPTIONS tab to open.
  - Under GUIDES, all options should be off.
  - Under LOCAL VIEW, turn all options off.
  - Under STARS, turn all options off.
  - Under SOLAR SYSTEM, turn on ASTEROIDS, turn on PLANETS-MOONS; turn all other options off.
  - Under CONSTELLATIONS, turn off all options.
  - Under DEEP SPACE, turn off all options.

- Click on the word OPTIONS to close this tab. At the moment, there may be few objects visible, as the program is now showing only those planets that are above the horizon from your viewing location.
- Change your viewing location so that you are looking down at the Sun and can see the orbits of asteroids and planets:
  - Select OPTIONS / VIEWING LOCATION from the drop-down menus at the top of the Starry Night window. The VIEWING LOCATION subwindow will appear.
  - At the top of the subwindow, click on the down-arrow to the right of the dialog box labeled VIEW FROM. Select STATIONARY LOCATION.
  - The box will change so that you can select a fixed location in space. These coordinates are centered on the Sun. Replace the numbers in the boxes labeled CARTESIAN COORDINATES, and enter the following values:
    - X: 0 au
    - Y: 0 au
    - Z: 15 au
- Click on the GO TO LOCATION button. The view will now shift to a location above the Sun. You can hit the space bar during the motion to speed up the process.
- Click on the FIND tab to open it.
  - Double-click on the word SUN in the NAME column. This will shift the angle of your gaze so that it is centered on the Sun.
  - Each planet listed has three checkboxes. The box to the left of the name will, if selected, turn on the name of the planet on the display. Turn on the names for Earth, Mars, and Jupiter.
  - The first box to the right of the name will, if selected, show the orbit of the planets. Turn on the orbits of Earth, Mars, and Jupiter.
- Click on the FIND tab to close.

- Zoom in or out using the + and − buttons at the top right of the Starry Night window. Adjust the zoom until the orbit of Jupiter fills the display but is still visible all around.

ACTIVITY 1 – KEPLER'S LAWS REVISITED

- Set the TIME FLOW RATE to two days. Turn on the flow of time with the (▶) button. Let it run for a while, then stop the flow of time with (■).

1. There are two planets that orbit inside Earth's orbit. These are not labeled. Which planets are they?
2. You will note that the orbital periods of the planets are larger if they are farther from the Sun. Which one of Kepler's laws describes this?

ACTIVITY 2 – ORIENTATION OF THE ORBITAL PLANES OF THE ASTEROIDS

- Using the dropdown menus at the top of the display, select OPTIONS / SOLAR SYSTEM / ASTEROIDS. A sub-window labeled ASTEROID OPTIONS will appear. Click and drag the slider bar labeled ASTEROID BRIGHTNESS until it is located to the right. Close the subwindow by clicking the **OK** button.
- Click on the FIND tab to open it.
- Open the list of dwarf planets by clicking on the + button next to the words DWARF PLANETS.
- Click on the ORBIT box to the right of Ceres to display its orbit.
- Open the list of asteroids by clicking on the + button next to the word ASTEROIDS.
- Click on the ORBIT boxes to display the orbits of the following asteroids: Hygiea, Pallas, and Vesta. (To make it easier to find these asteroids, click on the word NAME at the top of the column to alphabetize them.)
- Turn on the flow of time with the (▶) button. Let it run for a while, then stop the flow of time with (■).

3. Are these orbits all circular, like Earth's orbit? Are any even more elongated than the orbit of Mars? (You may need to click the orbits on and off one at a time to see the orbit shapes more clearly.)

- Double-click on the word SUN in the NAME column to center the view on the Sun. (It should already be centered, but this step may be necessary after selecting orbits above.)
- Close the FIND tab by clicking on the word FIND.
- Turn on the flow of time with the (▶) button.
- While the time is flowing, change the viewing location so that you are looking toward the Sun in the plane of Earth's orbit:

- Select OPTIONS / VIEWING LOCATION from the drop-down menus at the top of the Starry Night window. The VIEWING LOCATION subwindow will appear.
- At the top of the subwindow, click on the down-arrow to the right of the dialog box labeled VIEW FROM. Select STATIONARY LOCATION.
- The box will change so that you can select a fixed location in space. These coordinates are centered on the Sun. Replace the numbers in the boxes labeled CARTESIAN COORDINATES, and enter the following values:
  - X: 15 au
  - Y: 0 au
  - Z: 0 au
- Click on the GO TO LOCATION button. The view will now shift to a location in the plane of the Solar System.

4. Are the orbits of the asteroids all exactly in the plane of the Solar System or do they have significant tilts?

ACTIVITY 3 – ASTEROIDS THAT PASS NEAR EARTH

- Stop the flow of time with the (■) button.
- Turn off the dots representing the asteroids. Using the dropdown menus at the top of the display, select OPTIONS / SOLAR SYSTEM / ASTEROIDS. A subwindow labeled ASTEROID options will appear. Click and drag the slider bar labeled ASTEROID BRIGHTNESS until it is located to the left. Close the subwindow by clicking the **OK** button.
- Set the TIME and DATE so that it reads January 1, 1996.
- Shift back to a viewpoint that looks down on the Solar System.
- Select OPTIONS / VIEWING LOCATION from the dropdown menus at the top of the Starry Night window. The VIEWING LOCATION subwindow will appear.
- At the top of the subwindow, click on the down-arrow to the right of the dialog box labeled VIEW FROM. Select STATIONARY LOCATION.
- The box will change so that you can select a fixed location in space. These coordinates are centered on the Sun. Replace the numbers in the boxes labeled CARTESIAN COORDINATES, and enter the following values:
  - X: 0 au
  - Y: 0 au
  - Z: 5 au
- Turn off the orbits and labels for Mars and Jupiter. Turn on the label and orbit for Earth.
- Under ASTEROIDS, turn off the orbits for the asteroids selected in the previous activity.
- If your version of Starry Night has an asteroid named Bacchus in the list, turn on the orbit and name for Bacchus.

- If Bacchus is not present, you can create it using the following procedure:
  - Using the dropdown menus at the top of the Starry Night window, select FILE / NEW ASTEROID ORBITING SUN. A subwindow labeled ASTEROID: UNTITLED will appear.
  - There will be a dialog box at the top that contains the word UNTITLED. Replace UNTITLED with **Bacchus**.
  - The display will show a series of slider bars under a tab called ORBITAL ELEMENTS. To the right of each slider bar is a dialog box in which you can enter numbers manually. Enter the following numbers:
    - Mean distance (a): 1.07795
    - Eccentricity (e): 0.39494
    - Inclination (i): 9.4347
    - Ascending node: 33.17596
    - Arg[ument] of pericenter (w): 55.202
    - Mean anomaly (L): 191.126
    - Epoch: 2454000.5 (Julian day)
  - Close the subwindow by clicking on the × at the top right corner. You will be asked whether you want to save the new information. Select SAVE.
- Now find the entry under the USER CREATED list called Bacchus. Click the boxes to the left and to the right of the name Bacchus. This will show both the orbit of Bacchus and its position.
- Set the TIME FLOW RATE to six hours.
- Start the flow of time with the (▶) button. You will notice that Bacchus and Earth apparently came close together in late March.
- The fact that you are working in this exercise, means that Bacchus evidently did not strike Earth in 1996!

5. Advance a hypothesis about the orbit of Bacchus that would explain why it did not collide with Earth.

- Shift back to a viewpoint that is centered on the plane of the Solar System. Select OPTIONS / VIEWING LOCATION from the dropdown menus at the top of the Starry Night window. The VIEWING LOCATION subwindow will appear.
  - At the top of the subwindow, click on the down-arrow to the right of the dialog box labeled VIEW FROM. Select STATIONARY LOCATION.
  - The box will change so that you can select a fixed location in space. These coordinates are centered on the Sun. Replace the numbers in the boxes labeled CARTESIAN COORDINATES, and enter the following values:
    - X: -3 au
    - Y: -3 au
    - Z: 0 au
  - Click on the GO TO LOCATION button. The view will now shift to a location where you can see the orientation of Earth's and Bacchus' orbits. You can hit the space bar during the motion to speed up the process. Other planets, such as Jupiter, may also be in the view.
- Using the TIME FLOW buttons, run time forward or backward to explore the passage of Bacchus near Earth. The minimum separation was about 15 million km.

6. Was your hypothesis correct? Why did Bacchus miss Earth?
7. Can you speculate as to why the orbits of the asteroids differ from those of the planets?

Name: _____

Class/Section: _____

Starry Night Student Exercise – Answer Sheet
Asteroids

1. There are two planets that orbit inside Earth's orbit. These are not labeled. Which planets are they?

   _____

2. You will note that the orbital periods of the planets are larger if they are farther from the Sun. Which one of Kepler's laws describes this?

   _____

3. Are these orbits all circular, like Earth's orbit? Are any more elongated than the orbit of Mars?

   _____

4. Are the orbits of the asteroids all exactly in the plane of the Solar System, or do they have significant tilts?

   _____

5. Advance a hypothesis about the orbit of Bacchus that would explain why it did not collide with Earth.

   _____

   _____

6. Was your hypothesis correct? Why did Bacchus miss Earth?

   _____

   _____

7. Can you speculate as to why the orbits of the asteroids differ from those of the planets?

   _____

   _____

The Magnitude Scale and Distances

GOAL

- To be able to compare distances to stars using brightness and luminosity information

READING

- Section 13.1 – Measuring the Distance, Brightness, and Luminosity of Stars (esp. Connections 13.1 – The Magnitude System)
- Appendix 6 – Astronomical Magnitudes

Finding distances to stars is one of the most challenging problems in astronomy. Generally speaking, the farther away a light source is moved, the dimmer it looks. But when comparing stars, the luminosity also needs to be considered, since all stars do not emit an equal amount of energy. Because of the relation between a star's brightness and its luminosity (how much energy it emits per second), we can determine its distance.

Astronomers commonly make use of a scale called *magnitudes* to describe the brightness and luminosity of a star. This scale is set up so that smaller (and even negative) numbers represent brighter or more luminous stars. A star's **apparent magnitude** represents how bright it appears in our night sky. A star's **absolute magnitude** represents how bright a star would look if it were seen from a distance of 10 parsecs (32.6 ly). Since all stars' absolute magnitudes are expressed as if they were at that distance, the number is directly related to how luminous it is. When comparing stars, larger magnitude numbers mean dimmer or less luminous stars, whereas

smaller or more negative numbers mean brighter or more luminous stars. These values can also be used to compare distances. A star with very similar apparent and absolute magnitudes must be close to 10 pc away. For stars that are much farther away there will be a larger difference in the magnitudes. In this lab we will gather magnitude information for the main stars of the constellation Hercules and use this to compare luminosities and distances.

SETUP

- Start Starry Night.
- Stop the flow of time with the STOP TIME button (■).
- Click the OPTIONS tab to open.
  - Under LOCAL VIEW, turn on LOCAL HORIZON. Turn all other options off.
  - Under SOLAR SYSTEM, turn off all options.
  - Under STARS, turn on STARS. All other options should be off.
  - Under CONSTELLATIONS, turn on BOUNDARIES, LABELS, and STICK FIGURES. All other options should be off.
- Click the OPTIONS tab again to close.
- Set the TIME FLOW RATE to two minutes.
- Aim your gaze toward the east point of the horizon with the E button, located toward the upper right of the Starry Night window.
- Run time forward until the constellation Hercules has risen. If it is already up, use the handgrab tool to center Hercules on your screen. Zoom in until Hercules takes up most of the observation window.

ACTIVITY 1 – APPARENT AND ABSOLUTE MAGNITUDES OF STARS IN HERCULES

- Point at one of the bright stars in Hercules connected with the stick figure. The cursor will change to the heads-up display shape (▤).
- Click on the star. The name of the star will appear next to it.
- With the cursor at the same position (it has a ▤ shape), right-click on the star. Select the SHOW INFO option.
- The information on this star will appear on the left. Look at the information contained under OTHER DATA. If there is a + next to the words OTHER DATA, click on it and the OTHER DATA subwindow will expand to show several facts about this star.
- In the table on the answer sheet, write down the apparent magnitude and absolute magnitude for the star you have selected. The apparent magnitude is the sixth datum from the bottom. The absolute magnitude is the fifth datum from the bottom.
- Proceed to the other stars along the stick figure for Hercules.

1. Record the apparent magnitude and absolute magnitude for each star given in the table on your answer sheet, but do not fill in the BRIGHTNESS and LUMINOSITY RANKING columns yet.

ACTIVITY 2 – USING THE MAGNITUDE SCALE

- On your answer sheet, you should now have apparent and absolute magnitudes for 10 of the bright stars in Hercules. You will now use the information to rank the stars in two ways. First you will use the apparent magnitude numbers to rank the stars from brightest to dimmest. Then you will use the absolute magnitude numbers to rank their true luminosities, again from greatest to least. Remember that the magnitude scale is a reversed scale, referring to the reading at the beginning of this exercise or in the textbook to guide your rankings.

2. In the column labeled BRIGHTNESS RANKING, rank the stars from brightest to dimmest. If two stars have the same brightness, give them the same ranking.
3. In the column labeled LUMINOSITY RANKING, rank the stars from most to least luminous. If two stars have the same luminosity, give them the same ranking.
4. Are the brightest stars necessarily the most luminous? Explain why this may not be the case.

ACTIVITY 3 – COMPARING DISTANCES USING MAGNITUDES

5. Find two stars in the table with the same apparent magnitude. Which star is farther away? Explain how you can tell.
6. Find two stars in the table with the same absolute magnitude. Which one is farther away? Explain how you can tell.
7. Is the brightest star necessarily the closest? Explain why this may not be the case.
8. By comparing the apparent and absolute magnitudes, find which star is closest and which star is farthest. Explain your reasoning.

Name: _____

Class/Section: _____

Starry Night Student Exercise – Answer Sheet (continues on back)
The Magnitude Scale and Distances

1. Apparent and absolute magnitudes of bright stars in Hercules:

Star	Apparent magnitude	Absolute magnitude	Brightness ranking	Luminosity ranking
Epsilon Herculis				
Eta Herculis				
Gamma Herculis				
Iota Herculis				
Kornephoros				
Lambda Herculis				
Mu Herculis				
Omicron Herculis				
Phi Herculis				
Pi Herculis				
Rasalgethi				
Rho Herculis				
Sarin				
Sigma Herculis				
Tau Herculis				
Theta Herculis				
Xi Herculis				
Zeta Herculis				

2. In the column labeled BRIGHTNESS RANKING, rank the stars from brightest to dimmest. If two stars have the same brightness, give them the same ranking.

3. In the column labeled LUMINOSITY RANKING, rank the stars from most to least luminous. If two stars have the same luminosity, give them the same ranking.

4. Are the brightest stars necessarily the most luminous? Explain why this may not be the case.

_____

_____

5. Find two stars in the table with the same apparent magnitude. Which one is farther away? Explain how you can tell.

_____

_____

6. Find two stars in the table with the same absolute magnitude. Which one is farther away? Explain how you can tell.

_____

_____

7. Is the brightest star necessarily the closest? Explain why this may not be the case.

_____

_____

8. By comparing the apparent and absolute magnitudes, find which star is closest and which star is farthest. Explain your reasoning.

_____

_____

Stars and the H-R Diagram

GOAL

- To gather data on nearby stars and on bright stars, and to reproduce the results shown in Figure 13.16 of the text

READING

- Section 13.4 – The H-R Diagram Is the Key to Understanding Stars

The **Hertzsprung-Russell diagram** (or H-R diagram) is a fundamental tool for understanding the structure and evolution of stars. It is a plot of the luminosity of stars against their temperatures. In this exercise, we express the temperature using the Kelvin scale. The luminosity is given with respect to the solar luminosity; a star with luminosity = 100 suns, for example, emits 100 times as much energy each second as does the Sun.

In order to find which stars are most common, we look near the Sun and sample all stars in a specified volume of space. When we do so, we find that most stars are found near the **main sequence** on the H-R diagram. Along the main sequence, there is a well-defined **luminosity-temperature relationship**: stars run from hot/high luminosity to cool/low luminosity. Most stars in the local volume of space are cooler and less luminous than the Sun.

In contrast, most of the bright stars in the sky, such as those that make up the familiar constellations, are very luminous. As a result, they can be fairly bright even when located at long distances from the Sun.

SETUP

- Start Starry Night.
- Stop the flow of time with the STOP TIME button (■).
- Click the OPTIONS tab to open.
  - Under GUIDES, turn all options off.
  - Under LOCAL VIEW, turn on LOCAL HORIZON. Turn all other options off.
  - Under SOLAR SYSTEM, turn off all options.
  - Under STARS, turn on STARS; turn off MILKY WAY.
  - Under CONSTELLATIONS, turn on BOUNDARIES, LABELS, and STICK FIGURES. All other options should be off.
- Click the OPTIONS tab again to close.
- Set the TIME FLOW RATE to two minutes.
- Aim your gaze toward the east point of the horizon with the E button, located toward the upper right of the Starry Night window.
- Run time forward until the constellation Hercules has risen. If it is already up, use the handgrab tool to center Hercules on your screen. Zoom in until Hercules takes up most of the observation window.

ACTIVITY 1 – LUMINOSITIES AND TEMPERATURES OF BRIGHT STARS IN HERCULES

- Point at one of the bright stars in Hercules connected with the stick figure. The cursor will change to the heads-up display shape (▤).
- Click on the star. The name of the star will appear next to it.

- With the cursor at the same position (it has a ▤ shape), right-click on the star. Select the Show Info option.
- The information on this star will appear on the left. Look at the information contained under Other Data. If there is a + next to the words Other Data, click on it and the Other Data subwindow will expand to show several facts about this star.
- Write down the name of the star, the temperature, and the luminosity. The temperature is the third datum from the bottom. The luminosity is the bottom number. This may not be known for all stars.
- Proceed to the other stars along the stick figure for Hercules.

1. Transfer the luminosity and temperature for each bright star in Hercules to the table on your answer sheet.

ACTIVITY 2 – LUMINOSITIES AND TEMPERATURES OF NEARBY STARS

- Under the Options tab, turn off Local Horizon under Local View. Close the Options tab. This will allow you to see stars in all directions, even if they are below the horizon.
- Click on the Find tab to the left to open it.
- At the top of the window, find the dialog box that is labeled Search All Databases. You will be typing in several names of stars, then looking up the luminosity and temperature of each.
- For each star in the table on the next page, type in the name exactly as it appears in the left column. After entering the name, hit the Enter (or Return) key. The view will shift to the location of that star. You can hit the space bar while the motion is happening to speed up the process.

- The display will now show the star you have selected. Zoom in using the + key at the upper right until there are only a few stars on the display other than the marked star. Many of the stars are faint and are near bright stars, so you have to zoom in to make sure you can see the correct star.
- Using the procedures above, move the cursor over the star until it changes to the heads-up display shape (▤). Right-click on the star and select Show Info from the dropdown menu. Write down the temperature and the luminosity. Alternatively, right-click on the name of the object in the list of items found and select Show Info.
- Click on the Find tab again and repeat the process for the other stars in the table on the next page. You can leave the zoom the way it is between stars.

2. Here is the list of stars you will be looking for: HIP71683, HIP71681, HIP54035, HIP32349, HIP16537, HIP37279, HIP108870, and HIP439. These are among the nearest stars to the Sun. Also include the Sun, which has a luminosity of 1 and a temperature of 5,770 K.

ACTIVITY 3 – THE H-R DIAGRAM

3. On your answer sheet, you should now have tables of the luminosity and temperature for (a) several bright stars in the sky and (b) several nearby stars. Plot these data on an H-R diagram, also supplied on the answer sheet. Use different symbols for each group of stars. Include the information for the Sun in the plot for nearby stars.
4. Are most bright stars near the main sequence? What about the nearby stars?
5. Which sample of stars do you think is more representative of "average" stars in the galaxy? Why?

Name: _____

Class/Section: _____

Starry Night Student Exercise – Answer Sheet (continues on back)
Stars and the H-R Diagram

1. Luminosity and temperature of bright stars in Hercules:

Star	Luminosity	Temperature
Epsilon Herculis		
Eta Herculis		
Gamma Herculis		
Iota Herculis		
Kornephoros		
Lambda Herculis		
Mu Herculis		
Omicron Herculis		
Phi Herculis		
Pi Herculis		
Rasalgethi		
Rho Herculis		
Sarin		
Sigma Herculis		
Tau Herculis		
Theta Herculis		
Xi Herculis		
Zeta Herculis		

2. Luminosity and temperature of nearby stars:

Star	Luminosity	Temperature
Sun	1	5,770
HIP71683	1.8	5,715
HIP71681		
HIP54035		
HIP32349		
HIP16537		
HIP37279		
HIP108870		
HIP439		

3. H-R diagram of luminous and nearby stars (use a different symbol for each group):

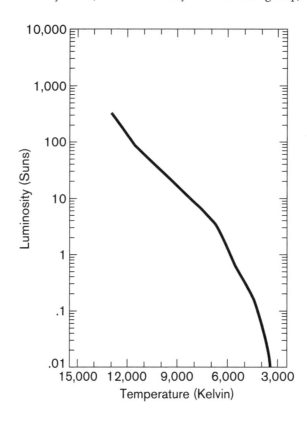

4. Are most bright stars near the main sequence? What about the nearby stars?

_____

_____

5. Which sample of stars do you think is more representative of "average" stars in the galaxy? Why?

_____

_____

Nebulae: The Birth and Death of Stars

GOAL

- To differentiate between the types of nebulae that give birth to stars and those that arise from the death of stars

READING

- Section 15.1 – The Interstellar Medium
- Section 16.4 – The Low-Mass Star Enters the Last Stages of Its Evolution

Nebulae are among the most beautiful and photogenic objects in space. Most basically, a **nebula** is a cloud of gas in space. In spiral galaxies like our Milky Way, clouds of gas are very common, making up a large part of the spiral arms. These clouds are generally dark and diffuse. It is when a cloud is lit up that it makes a more colorful showing. These colorful, showy nebulae come in a variety of types, forms, and colors, each arising from a different process. In this lab we will be investigating **emission nebulae** (also called H II regions) that give birth to stars as well as **planetary nebulae** that are the result of the end stages of the life of low-mass stars like our Sun.

SETUP

- Start Starry Night.
- Stop the flow of time with the STOP TIME button (■).
- Click the OPTIONS tab to open.
- Under LOCAL VIEW, turn all options off.
- Under SOLAR SYSTEM, turn on PLANETS-MOONS. Turn all other options off.

- Under STARS, turn on MILKY WAY and STARS. All other options should be off.
- Under DEEP SPACE, turn on BRIGHT NGC OBJECTS and MESSIER OBJECTS. Turn all other options off.
- Using dropdown menus, select OPTIONS / STARS / MILKY WAY. A subwindow labeled MILKY WAY OPTIONS will open. In this subwindow, click and slide the BRIGHTNESS slider bar all the way to the right. Click OK to close this subwindow.

ACTIVITY 1 – EMISSION NEBULAE

- Click on the FIND tab to open.
- You will be observing four different emission nebulae: North America Nebula, Eagle Nebula, Omega Nebula, and Lagoon Nebula. For each one, type the full name into the dialog box and press ENTER. The view will automatically change to center the nebula. You can press the space bar to quickly change your view directly to the nebula.
- Use the + and – keys at the top right of the screen to zoom in to display the nebula's image clearly.
- Repeat the procedure for each of the four nebulae, observing each one and noting any features such as shape, size, color, and so on.

1. What color do emission nebulae have in common? Why might they all share this particular color?
2. Do emission nebulae encompass one star or many stars?
3. Describe the locations of these four emission nebulae relative to the Milky Way.

ACTIVITY 2 – PLANETARY NEBULAE

- Click on the OPTIONS tab to open.
  - Under DEEP SPACE, turn on HUBBLE IMAGES. Leave all other options as they were.
- Click the FIND tab to open.
- In the FIND tab dialog box, erase what is there and type in **M57** and press ENTER. This will center your view on the Ring Nebula, a good example of a typical planetary nebula.
- Use the + and – keys at the top right of the screen to zoom in to display the nebula's image clearly.
- Repeat the procedure above to use the FIND tab to search for and observe four other nebulae, typing in their names as follows: Dumbbell, Blue Snowball, Blinking Planetary, Spirograph Nebula. As in the previous activity, observe each one and note any features such as shape, size, color, and so on.

4. Describe the basic shape of planetary nebula.
5. How many stars does each of these planetary nebula contain?
6. Are planetary nebulae larger or smaller than emission nebulae? How can you tell?

7. Compare the colors of the planetary nebulae you have observed with the color of the emission nebulae you observed earlier. What could explain this difference?

- Telescopes in the 1700s and 1800s, when these nebulae were first being identified and named, had a relatively small magnification compared with today's telescopes. To simulate this, use the + and – keys at the top right of the screen to set the zoom to 1° × 1°.
- Click the FIND tab to open.
- In the dialog box, type **Jupiter** and press ENTER to center your view on this planet. It will be the small disk in the center of your screen. Do not zoom in.
- Now clear the dialog box and type **M57** and press ENTER. This will center your view on the Ring Nebula again.

8. After observing both Jupiter and the Ring Nebula at the same low magnification, can you come up with an explanation for why this type of nebula is called a "planetary" nebula?

Name: _____

Class/Section: _____

Starry Night Student Exercise – Answer Sheet
Nebulae: The Birth and Death of Stars

1. What color do emission nebulae have in common? Why might they all share this particular color?

   _____

   _____

2. Do emission nebulae encompass one star or many stars?

   _____

3. Describe the locations of these four emission nebulae relative to the Milky Way.

   _____

   _____

4. Describe the basic shape of planetary nebula.

   _____

5. How many stars does each of these planetary nebula contain?

   _____

6. Are planetary nebulae larger or smaller than emission nebulae? How can you tell?

   _____

7. Compare the colors of the planetary nebulae you have observed with the color of the emission nebulae you observed earlier. What could explain this difference?

   _____

   _____

8. After observing both Jupiter and the Ring Nebula at the same low magnification, can you come up with an explanation for why this type of nebula is called a "planetary" nebula?

   _____

   _____

Pulsars and Supernova Remnants

GOAL

- To investigate the products of supernova explosions

READING

- Section 16.5 – Binary Star Evolution
- Section 17.3 – The Spectacle and Legacy
  of Supernovae

Supernova explosions are the very violent deaths of certain types of stars. Tremendous energy is released during such an event, which will briefly outshine an entire galaxy, making them very visible. There are two basic types of supernovae, depending on the original star. When a white dwarf star (the remnant of a low-mass main-sequence star) is pushed over a certain mass limit, it will detonate in a **type Ia supernova**, which completely destroys the star. When a high-mass main-sequence star undergoes core collapse, it explodes in a **type II supernova** that can leave behind the exposed collapsed core. In this case, the exposed core is a superdense compact ball of solid neutrons called a **neutron star**, which is typically the size of a city.

All supernovae leave behind an expanding cloud of debris called a **supernova remnant**. Astronomers generally view the remnants of these explosions, rather than observing the explosion directly. However, even this eventually fades from view as it diffuses itself into the surrounding space, distributing the elements created during the explosion into the galaxy to be incorporated into future generations of stars. Recall that the type II supernova also leaves behind a small, very dense neutron star.

The neutron stars created in type II supernovae will often be spinning at fantastic rates, emitting sweeping beams of radiation from their poles that we see as a **pulsar**. The location of a pulsar tells where a supernova once occurred, even if the remnant debris cloud is no longer visible. In this exercise we will observe the location and properties of the remnants of several of the more recent of these explosions.

SETUP

- Start Starry Night.
- Stop the flow of time with the STOP TIME button (■).
- Click the OPTIONS tab to open.
  - Under LOCAL VIEW, turn all options off.
  - Under SOLAR SYSTEM, turn all options off.
  - Under STARS, turn on MILKY WAY and PULSARS. All other options should be off.
  - Under CONSTELLATIONS, turn on BOUNDARIES and LABELS.
  - Under DEEP SPACE, turn on MESSIER OBJECTS. Turn all other options off.
- Click the OPTIONS tab again to close.
- Using the dropdown menus, select OPTIONS / STARS / MILKY WAY. A subwindow labeled MILKY WAY OPTIONS will open. In this subwindow, click and slide the BRIGHTNESS slider bar all the way to the right. Click OK to close this subwindow.
- Using the dropdown menus, select OPTIONS / STARS / PULSARS. A subwindow labeled PULSAR OPTIONS will open. In this subwindow, click and slide the NUMBER OF OBJECTS slider bar all the way to the right. Click OK to close this subwindow.

ACTIVITY 1 – PULSARS

- Use the – key at the top right of your screen to zoom out to near maximum view. Your screen should look like a circle with the locations of pulsars marked and the band of the Milky Way stretching across.
- Using the handgrab tool, swivel your view around to observe where pulsars lie in the sky relative to the Milky Way.

1. What do you notice about the location of most pulsars relative to the Milky Way? Can you think of a reason to explain this?
2. Which constellation contains the most pulsars? Is this constellation in or near the direction of the center of the Milky Way? Why would this direction contain more pulsars?

- Click on the FIND tab to open.
- At the top of the window, find the dialog box labeled SEARCH ALL DATABASES. Type in the name **M1** and press ENTER. The program will take a moment to search for the entry and will shift your view to center the location of M1. This is the remnant of a supernova, observed nearly 1,000 years ago, that is now called the Crab Nebula.
- Use the + key at the top right of your screen to zoom in to see the Crab Nebula clearly.
- At the center of the nebula is a mark representing a pulsar. Point at the marker. The cursor will change to the heads-up display shape (▤).
- Right-click on the pulsar to bring up a menu. Select the SHOW INFO option.
- The information for this pulsar will appear on the left. Look at the information contained in the top

section and find "barycentric per." This is the rotational period of the Crab Nebula pulsar.

3. What is the period of rotation of the Crab Nebula pulsar? How many times does it spin in one second?

ACTIVITY 2 – SUPERNOVA REMNANTS

- Click the OPTIONS tab to open.
  - Under DEEP SPACE, turn on BRIGHT NGC OBJECTS and CHANDRA IMAGES.
  - Under OTHER, turn on SUPERNOVA REMNANTS.
- Click the FIND tab to open.
- Find and observe each of the supernova remnants given in the table on your answer sheet. Try to determine if each could be a type Ia or type II supernova and record this in the table, along with a brief justification for your decision. Remember that we can distinguish between the two depending on what they leave behind. In the cases of remnants with pulsars, the images provided by the program may not be exactly lined up with the correct position of the pulsar.

4. Determine the type of supernova remnants listed in the table and provide a brief justification.
5. Compare the size of the Cygnus Loop (also known as the Veil Nebula) with the other remnants. What does this tell you about its relative age?
6. Do all of these supernova remnants have pulsars in their center? Why or why not?
7. Do all pulsars have supernova remnants surrounding them? Why or why not?

Name: _____

Class/Section: _____

Starry Night Student Exercise – Answer Sheet (continues on back)
Pulsars and Supernova Remnants

1. What do you notice about the location of most pulsars relative to the Milky Way? Can you think of a reason to explain this?

   _____

   _____

2. Which constellation contains the most pulsars? Is this constellation in or near the direction of the center of the Milky Way? Why would this direction contain more pulsars?

   _____

   _____

3. What is the period of rotation of the Crab Nebula pulsar? How many times does it spin in one second?

   _____

4. Determine the type of supernova remnants listed in the table and provide a brief justification.

Supernova remnant	Type (Ia or II)	Justification
Crab Nebula		
Tycho		
Cassiopeia A		
Vela Pulsar		
B1509-58		
Cygnus Loop		

5. Compare the size of the Cygnus Loop (also known as the Veil Nebula) with the other remnants. What does this tell you about its relative age?

   _____

6. Do all of these supernova remnants have pulsars in their center? Why or why not?

   _____

   _____

7. Do all supernova remnants have pulsars in their center? Why or why not?

_____

_____

Galaxy Classification

GOAL

- To be able to classify galaxies based on their appearance

READING

- Section 20.1 – Galaxies Come in Many Types

In every direction in the sky, when we look beyond the nearby stars of our own Milky Way galaxy we see uncountable numbers of other galaxies. Edwin Hubble (after whom the Hubble space telescope was named) was the first to come up with a classification scheme by observing patterns in a large number of galaxies. Galaxies seem to fall into two basic shapes: spiral and elliptical. *Ellipticals* are classified by how elongated they are, from E0 to E5. *Spirals* are classified by how tight their arms are and how prominent their central bulge is, from Sa to Sd. A small number of galaxies have no regular pattern and are called *irregular galaxies*. In this lab you will observe many galaxies and learn how to tell the differing shapes apart.

SETUP

- Start Starry Night.
- Stop the flow of time with the STOP TIME button (■).
- Click the OPTIONS tab to open.
  - Under LOCAL VIEW, turn all options off.
  - Under SOLAR SYSTEM, turn all options off.
  - Under STARS, turn on MILKY WAY and STARS. All other options should be off.
  - Under DEEP SPACE, turn on BRIGHT NGC OBJECTS and MESSIER OBJECTS. Turn all other options off.

- Using the dropdown menus, select OPTIONS / STARS / MILKY WAY. A subwindow labeled MILKY WAY OPTIONS will open. In this subwindow, click and slide the BRIGHTNESS slider bar all the way to the right. Click OK to close this subwindow.

ACTIVITY 1 – SPIRAL GALAXIES

- Click on the FIND tab to open.
- In the dialog box, type in **M101** and press ENTER. The view will automatically change to center this galaxy. You can press the space bar to quickly change your view directly to the galaxy.
- Use the + and − keys at the top right of the screen to zoom in to display the galaxy's image clearly.

1. Describe the basic shape of M101 along with any color differences within the galaxy.
2. What could explain the color difference between the arms and the central core of a spiral galaxy? (Hint: Think about stars of differing colors.)

- In the FIND tab dialog box, erase what is there, type in **M31**, and press ENTER.

3. M31 (the Andromeda Galaxy) is another spiral galaxy like M101. What is different about our view of M31 that makes it look so different from M101?

- In the FIND tab dialog box, erase what is there, type in **NGC 4565**, and press ENTER.

4. What does this view of a spiral galaxy tell you about the overall shape of the arms and disk?

5. Estimate the thickness of the arms as a fraction of the diameter of the galaxy.

ACTIVITY 2 – ELLIPTICAL GALAXIES

- In the FIND tab dialog box, erase what is there and type in **M87**.
- In the list of results is an M87 with the ⊞ icon next to it. Click on the ⊞ icon to reveal several options.
- Click on the box next to the M87 that lists "Elliptical E0 Galaxy" in the KIND column.
- Click on the ▼ box to the left of the appropriate entry and select CENTER to go there.

6. Describe the basic shape of an elliptical galaxy.
7. What color does this galaxy appear to have? What does this tell you about the types of stars found in this galaxy?

ACTIVITY 3 - IRREGULAR GALAXIES

- In the FIND tab dialog box, erase what is there, type in **LMC**, and press ENTER.

8. The Large Magellenic Cloud (LMC) is a satellite irregular galaxy of the Milky Way. Does the LMC share characteristics of either spiral or elliptical galaxies?

ACTIVITY 4 – GALAXY CLASSIFICATION

9. Use the FIND tab to search for each of the galaxies in the table on your answer sheet. Classify each galaxy using Hubble's tuning fork diagram in Figure 20.3 as a guide.

- Click the FIND tab to close.
- Use the – key at top right to zoom out to the maximum view. At this point you should see the band of our Milky Way stretching across the sky.

10. When we look at our own Milky Way galaxy, our position inside of it makes it challenging to determine its true shape. How does our view of the Milky Way give evidence of which type of galaxy it is?

Name: _____

Class/Section: _____

Starry Night Student Exercise – Answer Sheet (continues on back)
Galaxy Classification

1. Describe the basic shape of M101 along with any color differences within the galaxy.

   _____

   _____

2. What could explain the color difference between the arms and the central core of a spiral galaxy? (Hint: Think about stars of differing colors.)

   _____

   _____

3. M31 (the Andromeda Galaxy) is another spiral galaxy like M101. What is different about our view of M31 that makes it look so different from M101?

   _____

   _____

4. What does this view of a spiral galaxy tell you about the overall shape of the arms and disk?

   _____

5. Estimate the thickness of the arms as a fraction of the diameter of the galaxy.

   _____

6. Describe the basic shape of an elliptical galaxy.

   _____

7. What color does this galaxy appear to have? What does this tell you about the types of stars found in this galaxy?

   _____

   _____

8. What characteristics does the LMC share with either spiral or elliptical galaxies?

   _____

9. Table of galaxies in the Messier Catalog:

Galaxy	Galaxy type
LMC	Irr
M87	E0
M101	Sc
M32	
M51	
M59	
M60	
M63	
M64	
M65	
M74	
M82	
M85	
M90	
M104	
M105	
SMC	

10. How does our view of the Milky Way give evidence of which type of galaxy it is?

_____

_____

_____

Quasars and Active Galaxies

GOAL

- To investigate the distribution and luminosities of quasars and active galaxies

READING

- Section 20.4 – Most Galaxies Have a Supermassive Black Hole at the Center

Among the vast number of galaxies surrounding us in the universe are some that emit incredible amounts of energy, called **active galactic nuclei (AGNs)**. The most luminous of these objects are called **quasars**, which is short for *quasi*-stell*ar* radio source. These were first detected as radio sources but are now known to correspond to very faint and distant galaxies. The luminosity of quasars is incredible, equaling the combined light of trillions of Sun-like stars, with all of the energy coming from a very compact source in the center of a galaxy.

Other types of AGNs are very similar, but less luminous. It is likely that these are all powered by large and active accretion disks surrounding supermassive black holes in the middle of galaxies that are undergoing some sort of activity like a galactic merger. The interaction drives material into the accretion disk, creating a beacon that can be seen over billions of light years. In this lab we will investigate some of the visible properties of quasars and active galaxies.

SETUP

- Start Starry Night.
- Stop the flow of time with the STOP TIME button (■).

- Click the OPTIONS tab to open.
  - Under LOCAL VIEW, turn all options off.
  - Under SOLAR SYSTEM, turn all options off.
  - Under STARS, turn on MILKY WAY and STARS. All other options should be off.
  - Under DEEP SPACE, turn on QUASARS. Turn all other options off.
  - Click the ⊞ button next to QUASARS to display other options. Turn off ACTIVE GALAXY and BL LAC OBJECT, leave QUASARS on.
- Click the OPTIONS tab again to close.
- Using the dropdown menus, select OPTIONS / STARS / MILKY WAY. A subwindow labeled MILKY WAY OPTIONS will open. In this subwindow, click and slide the BRIGHTNESS slider bar all the way to the right. Click OK to close this subwindow.
- Using the dropdown menus, select OPTIONS / DEEP SPACE / QUASARS. A subwindow labeled QUASAR OPTIONS will open. In this subwindow, click and slide the NUMBER OF OBJECTS slider bar all the way to the right. Click OK to close this subwindow.

ACTIVITY 1 – QUASARS

- Use the – key at the top right of your screen to zoom out to near maximum view. Your screen should look like a circle with the locations of quasars marked with crosses and the band of the Milky Way stretching across.
- Using the handgrab tool, swivel your view around to observe where quasars lie in the sky relative to the Milky Way.

1. Describe the distribution of quasars relative to the Milky Way. Does this imply the quasars are inside or outside of the Milky Way?

- Click on the OPTIONS tab to open.
- Under CONSTELLATIONS, turn on BOUNDARIES, LABELS, and STICK FIGURES.
- Click the OPTIONS tab again to close.
- Using the handgrab tool, scroll your view around the sky until you find the constellation Ursa Major. Using the + and – keys at the top right of your screen, zoom in until as much of the constellation as possible fills your screen.

2. Choose four of the seven quasars in Ursa Major and record their names, redshift, apparent magnitude, and absolute magnitude in the table on your answer sheet.
3. Our Sun has an apparent magnitude of about −26.7 in our daytime sky. Compare the absolute magnitudes of these quasars (how bright they would look at 10 pc) with how bright the Sun looks in our daytime sky. (Note: You would not want to be 10 pc from a quasar in order to observe this!)

- In a previous exercise (The Magnitude Scale and Distances), we used the apparent and absolute magnitudes of stars to estimate their distances. The same thing can be done to estimate the distances to quasars. Here are a few examples of the magnitudes and distances of objects within our Milky Way:

Name	Apparent magnitude	Absolute magnitude	Distance
Sirius	0.00	0.55	9 ly
Deneb	1.25	−6.95	1,400 ly
M13	7.0	−8.7	25,000 ly

4. Using the trends seen in the examples given above, compare the apparent and absolute magnitudes of the quasars you have chosen and use this to say something about the distance to these objects. Explain your reasoning.

ACTIVITY 2 – ACTIVE GALAXIES

- Click the OPTIONS tab to open.
  - Under DEEP SPACE / QUASARS, turn on ACTIVE GALAXY and turn off QUASARS.
- Click the OPTIONS tab again to close.

5. Find the trapezoid shape that represents the bowl of the Big Dipper in Ursa Major. Choose four active galaxies within that part of the constellation and record their names, redshift, apparent magnitude, and absolute magnitude in the table on your answer sheet.
6. Compare the absolute magnitudes of these galaxies with those of the quasars you previously recorded. Are active galaxies more or less intrinsically bright than quasars? Explain your reasoning.
7. As we did in question 4 above, compare the apparent and absolute magnitudes of these AGNs to make a distance estimate. Are AGNs generally closer or farther from us than quasars? Explain your reasoning.
8. Remember that according to Hubble's law, the redshift of distant galaxies tells us about how far away they are. Which type of object generally has a greater redshift, quasars or AGNs? What is the relation between redshift and your distance estimates for these two types of objects?

Name: _____

Class/Section: _____

Starry Night Student Exercise – Answer Sheet (continues on back)
Quasars and Active Galaxies

1. Describe the distribution of quasars relative to the Milky Way. Does this imply the quasars are inside or outside of the Milky Way?

   _____

   _____

   _____

2. Choose four of the seven quasars in Ursa Major and record their names, redshift, apparent magnitude, and absolute magnitude in the table. Calculate the average values of both the absolute and apparent magnitudes.

Quasar	Redshift	Apparent magnitude	Absolute magnitude
Average of magnitudes			

3. Our Sun has an apparent magnitude of about −26.7 in our daytime sky. Compare the absolute magnitudes of these quasars (how bright they would look at 10 pc) with how bright the Sun looks in our daytime sky.

   _____

   _____

4. Compare the apparent and absolute magnitudes of the quasars you have chosen and use this to say something about the distance to these objects. Explain your reasoning.

   _____

   _____

5. Choose four active galaxies within that part of the constellation and record their names, redshift, apparent magnitude, and absolute magnitude in the table on your answer sheet. Calculate the average values of both the absolute and apparent magnitudes.

Active galaxy	Redshift	Apparent magnitude	Absolute magnitude
Average of magnitudes			

6. Compare the absolute magnitudes of these galaxies with those of the quasars you previously recorded. Are active galaxies more or less intrinsically bright than quasars? Explain your reasoning.

7. As we did in question 4 above, compare the apparent and absolute magnitudes of these AGNs to make a distance estimate. Are AGNs generally closer or farther from us than quasars? Explain your reasoning.

8. Remember that according to Hubble's law, the redshift of distant galaxies tells us about how far away they are. Which type of object generally has a greater redshift, quasars or AGNs? What is the relation between redshift and your distance estimates for these two types of objects?

Views of the Milky Way

GOAL

- To investigate the appearance of the Milky Way in multiple wavelength views

READING

- Section 5.2 – Light Is an Electromagnetic Wave
- Section 21.1 – Measuring the Shape and Size of the Milky Way

Visible light is one type of **electromagnetic radiation**, with a very specific range of wavelengths, representing just a small portion of the entire electromagnetic spectrum. There are many other types of radiation that we cannot see that differ from visible light only in their wavelength. These types in order of longest to shortest wavelengths are radio, microwave, infrared, visible, ultraviolet, x-rays, and gamma rays. Types of radiation that have longer wavelengths than visible light (such as infrared and radio) are less energetic than visible light and are produced by cool objects. Types of radiation that have shorter wavelengths than visible light (such as ultraviolet, x-rays, and gamma rays) are more energetic than visible light and are produced by hotter objects.

Until the mid-20th century, our view of the universe was limited to only what we could see with our eyes—visible light. But today we have constructed telescopes that are designed to detect all other wavelengths of radiation, which give us a new view of our surroundings. Starry Night has a compilation of views of the Milky Way taken by many of these telescopes, such as the Chandra X-ray Observatory, the Spitzer infrared telescope, and many others that observe all regions of the EM spectrum. In this exercise we will investigate the Milky Way in a variety of different radiations to learn more about its nature and structure.

SETUP

- Start Starry Night.
- Stop the flow of time with the STOP TIME button (■).
- Click the OPTIONS tab to open.
  - Under GUIDES/ECLIPTIC GUIDES, turn on THE ECLIPTIC to show the plane of our Solar System. Turn all other options off.
  - Under LOCAL VIEW, turn all options off.
  - Under SOLAR SYSTEM, turn all options off.
  - Under STARS, turn on MILKY WAY and STARS. All other options should be off.
  - Under CONSTELLATIONS, turn on BOUNDARIES and LABELS.
  - Under DEEP SPACE, turn all options off.
- Click the OPTIONS tab again to close.

ACTIVITY 1 – THE MILKY WAY IN VISIBLE LIGHT

- Using the dropdown menus at the top of the display, select OPTIONS / STARS / MILKY WAY. A subwindow labeled MILKY WAY OPTIONS will appear.
- Click and drag the BRIGHTNESS slider bar all the way to the right.
- The WAVELENGTH should be set at its default of VISIBLE SPECTRUM.
- Click OK to close this subwindow.
- Use the handgrab tool to swivel your view around the sky to observe the appearance of the Milky Way.

1. Describe the appearance of the Milky Way.
2. What does the appearance of the Milky Way tell us about the overall shape of the galaxy?

3. Toward which constellation do you think is the center of the galaxy? Explain why you chose this direction.
4. Is the plane of the Solar System lined up with the plane of the Milky Way, or are the two planes at a large angle to each other?

ACTIVITY 2 – THE CENTER OF THE MILKY WAY IN OTHER WAVELENGTHS

- Using the dropdown menus at the top of the display, select OPTIONS / STARS / MILKY WAY. A subwindow labeled MILKY WAY OPTIONS will appear.
- Click on the (▼) button to the right of the WAVELENGTH box. This will bring down a list of other wavelength views of the Milky Way. Select NEAR-INFRARED. This will show you a longer wavelength radiation.
- Click OK to close this subwindow.

5. In which direction is IR radiation brightest? Does this match with your answer to question 3?

- Using the same instructions as above, observe the center of the Milky Way in all wavelength views.

6. Which types of radiation give us the best view of the center of the Milky Way?
7. Which types of radiation do we not see from the center of the Milky Way?

8. What does this tell you about the conditions at the center of our galaxy?

ACTIVITY 3 – OTHER PARTS OF THE MILKY WAY

- Using the same instructions as above, set the wavelength view to X-RAYS.
- Click on the OPTIONS tab to open.
  ∘ Under DEEP SPACE, turn on CHANDRA IMAGES and LABELS
- Click the OPTIONS tab again to close.

9. Looking around the sky, you'll see some very bright sources of x-rays that are not in the direction of the center of the Milky Way. Find two bright x-ray sources that have labeled names, and identify the constellations they are located in.
10. Observe the same two objects in all other wavelengths. Which other wavelengths do these two objects emit? What does this tell you about the nature of these two objects?

- Click on the OPTIONS tab to open.
- Under OTHER, turn on SUPERNOVA REMNANTS. These are clouds of debris left behind when stars explode violently, and they emit a great deal of high-energy radiation.
- Click on the OPTIONS tab again to close.

11. Describe the distribution of supernova remnants and the relation between this distribution and the gamma-ray view of the Milky Way.

Name: _____

Class/Section: _____

Starry Night Student Exercise – Answer Sheet (continues on back)
Views of the Milky Way

1. Describe the appearance of the Milky Way.

_____

_____

2. What does the appearance of the Milky Way tell us about the overall shape of the galaxy?

_____

3. Toward which constellation do you think is the center of the galaxy? Explain why you chose this direction.

_____

_____

4. Is the plane of the Solar System lined up with the plane of the Milky Way, or are the two planes at a large angle to each other?

_____

5. In which direction is IR radiation brightest? Does this match with your answer to question 3?

_____

6. Which types of radiation give us the best view of the center of the Milky Way?

_____

7. Which types of radiation do we not see from the center of the Milky Way?

_____

8. What does this tell you about the conditions at the center of our galaxy?

_____

9. Find two bright x-ray sources that have labeled names, and identify the constellations they are located in.

_____

_____

10. Observe the same two objects in all other wavelengths. Which other wavelengths do these two objects emit? What does this tell you about the nature of these two objects?

_____

_____

11. Describe the distribution of supernova remnants and the relation between this distribution and the gamma-ray view of the Milky Way.

_____

_____

Globular Clusters

GOAL

- To use the distribution of globular clusters in our galaxy to estimate the direction and distance to the center of the Milky Way

READING

- Section 21.1 – Measuring the Shape and Size of the Milky Way

In the early 20th century there was a debate about the nature of our galaxy. The presiding view was that since the Milky Way obviously surrounded our Solar System on all sides, our Solar System must be either at the center or somewhere very near to it. But it was noticed even in the early 1800s that globular clusters of stars appeared to be concentrated in one particular direction in the sky, rather than being evenly distributed in all directions, as would be expected if our Solar System were at the center of the galaxy.

If these **globular clusters**, which are compact clusters of millions of stars within a 1,000-light-year span, are not orbiting around the Solar System, then there must be another common center elsewhere in our galaxy. By determining the distances to these clusters and plotting their locations in space, astronomers were able to make a rough estimation of the direction and distance to the center of our galaxy. In this lab you are reproducing the type of work that astronomers do when researching a sample of objects. We will use Starry Night to find the distances to a selection of globular clusters in the direction of their highest concentration and make a plot of their distribution to find the location of the galactic center.

SETUP

- Start Starry Night.
- Stop the flow of time with the STOP TIME button (■).
- Click the OPTIONS tab to open.
  - Under GUIDES / GALACTIC GUIDES, turn on the EQUATOR and GRID. Turn all other options off.
  - Under LOCAL VIEW, turn all options off.
  - Under SOLAR SYSTEM, turn all options off.
  - Under STARS, turn on MILKY WAY and GLOBULAR CLUSTERS. All other options should be off.
  - Under CONSTELLATIONS, turn on BOUNDARIES and LABELS.
  - Under DEEP SPACE, turn all options off.
- Click the OPTIONS tab again to close.
- Using the dropdown menus, select OPTIONS / STARS / MILKY WAY. A subwindow labeled MILKY WAY OPTIONS will open. In this subwindow, click and slide the BRIGHTNESS slider bar all the way to the right. Click OK to close this subwindow.

ACTIVITY 1 – THE DISTRIBUTION OF GLOBULAR CLUSTERS

- Using the handgrab tool, click on the screen and drag your view around, looking at the entire circle of the Milky Way and paying attention to the distribution of globular clusters, which are represented as small light green circles. Alternately, use the arrow keys on your keyboard to shift your view all the way around.

1. Describe the general distribution of globular clusters relative to the visible band of the Milky Way.

2. Toward which constellation(s) do there appear to be the most globular clusters?
3. Does the Milky Way look visibly different in the direction of the constellation you chose? If yes, explain how.

ACTIVITY 2 – THE DISTANCES TO GLOBULAR CLUSTERS

- Click on the FIND tab to open.
- Find the dialog box labeled SEARCH ALL DATABASES. You will be typing in the names of several globular clusters here, then finding distance and angle information for each. All of these clusters have been chosen from the dense grouping near the constellation Sagittarius.
- For each globular cluster in the table on the answer sheet, type in the name as it appears in the left column (such as NGC 5904) and press ENTER. The program will take a moment to search for the entry and will shift your view to the location of that cluster.
- The object name will appear in the FIND tab with a ⊞ to the left. Click on the ⊞ to expand.
- The cluster name will be shown one or two more times. Click the ☑ box to the left of the entry from the GLOBULAR CLUSTERS database. (You may have to expand the FIND tab window to see the database column.)
- Click on the INFO tab to bring up information for this cluster.
- Under INFO is an item labeled DISTANCE TO SUN (LY). Write this down in the table.
- Under POSITION IN SKY is an item labeled GALACTIC LATITUDE. This represents how many degrees a

cluster is located above or below the apparent plane of the Milky Way. Write down the first number (degrees) in the table.
- Click the FIND tab again and repeat the process for the other globular clusters in the table on the answer sheet.

4. Complete the table of information for each globular cluster.

ACTIVITY 3 – THE DISTANCE TO THE GALACTIC CENTER

5. On your answer sheet, you should now have a table of distance and angle information for many globular clusters that are grouped in the same direction in our sky. Using this information, plot each cluster on the diagram provided on the answer sheet. The diagram is a set of concentric, equally spaced semicircles with the Sun at their center. Each semicircle represents a distance of 5,000 ly. Galactic latitude is measured in degrees above or below the galactic plane, represented by the horizontal 0° line. It is generally easiest to find the distance first along this line, then follow the semicircles around to the proper angle. The first two globular clusters are plotted for you.
6. Mark the apparent center of your plotted distribution of clusters with a big X. This is a rough estimate of the position of the center of our galaxy. How many light-years is the Sun from the center of our galaxy, according to your estimate?
7. What is the distance to the center of the galaxy as given in the textbook? How does this compare with the value you found?

Name: _____

Class/Section: _____

Starry Night Student Exercise – Answer Sheet (continues on back)
Globular Clusters

1. Describe the general distribution of globular clusters relative to the visible band of the Milky Way.

   _____

   _____

2. Toward which constellation(s) does (do) there appear to be the most globular clusters?

   _____

3. Does the Milky Way look visibly different in the direction of the constellation you chose? If yes, explain how.

   _____

   _____

4. Table of distances and galactic latitude angles for many globular clusters:

Cluster	Distance (ly)	Latitude
NGC 7099	41,000	30°
NGC 5897	24,000	46°
NGC 5904		
NGC 6093		
NGC 6121		
NGC 6144		
NGC 6171		
NGC 6235		
NGC 6284		
NGC 6316		
NGC 6333		
NGC 6453		
NGC 6522		
NGC 6544		
NGC 6652		
NGC 6681		
NGC 6723		
NGC 6809		
NGC 6864		

5. Plot each of the clusters in the table above according to distance versus galactic latitude. Each semicircle represents 5,000 ly. The first two have been plotted for you.

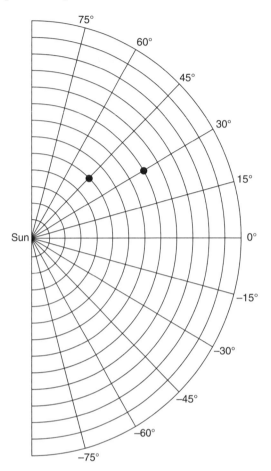

6. Mark the apparent center of your plotted distribution of clusters with a big X. This is a rough estimate of the position of the center of our galaxy. How many light-years is the Sun from the center of our galaxy, according to your estimate?

_____

7. What is the distance to the center of the galaxy as given in the textbook? How does this compare with the value you found?

_____

_____

The Neighborhood of the Sun

GOAL

- To illustrate the concept of high-velocity stars as samples of the local disk and halo of the Milky Way

READING

- Part of section 21.3 – The Halo Is More Than Globular Clusters

Studying the Milky Way in detail requires that we learn about stars at varying distances from the Sun. The presence of dust in the plane of the galaxy is a great complication, as it limits our view along the plane. Astronomers must therefore turn to stars that are near the Sun.

This is less of a disadvantage than it might first appear. Stars near the Sun might be members of the disk; as they orbit the Milky Way they spend most of their time near the plane of our galaxy. Local stars might also be members of the halo that just happen to be passing near the Sun.

We can find nearby stars in both groups by looking for **high-velocity stars**. These stars are found because their positions on the sky change quickly compared to other stars, although their motions are still slow in human terms. In this exercise, we identify some nearby high-velocity stars and demonstrate their apparent motion.

SETUP

- Start Starry Night.
- Stop the flow of time with the STOP TIME button (■).
- Click the OPTIONS tab to open.
  - Under GUIDES, turn all options off.

- Under LOCAL VIEW, turn all options off. In particular, make sure LOCAL HORIZON is unchecked – this will allow you to see stars in all directions.
  - Under SOLAR SYSTEM, turn all options off.
  - Under STARS, turn on STARS. Leave other options off.
  - Under CONSTELLATIONS, turn on BOUNDARIES, LABELS, and STICK FIGURES. Turn other options off.
- Click the FIND tab. In the dialog box type in the name **Rigil Kentaurus** but do not press ENTER.
- The name will appear in the list below with a ⊞ to the left. Click this box to expand the options. There will now be three entries with the same name.
- Double-click on the second entry and the view will shift to center on Rigil Kentaurus, also known as Alpha2 Centauri.
- If the box to the left of this entry is not checked, click in the box to check it. A pointer with the name should appear on the Starry Night window.
- In the dialog box, erase what is there and type in **Alpha2 Centauri** but do not press ENTER. Its name will appear in the list. Do NOT double-click on Alpha2 Centauri, but do click the left box so that this star is shown in the Starry Night display.
- Click on the FIND tab to close it.
- Using the mousewheel or the + button at the upper right of the display, zoom in until the value listed in the height is 4°.
- Set the TIME FLOW RATE to read 999 sidereal days.

ACTIVITY 1 – FINDING NEARBY STARS

- Pick any other reasonably bright star near the pair of selected stars. Run the cursor over it until the hand changes to the heads-up display shape (▤). Right-click and select CENTER. The view will shift slightly so that the star you selected is now centered.

- Start the flow of time. The view will change about 2.7 years per time step. This will make the proper motion of Rigil Kentaurus and Alpha2 Centauri visible. Run the time forward until about the year A.D. 3500; then stop the flow of time.

1. Consider driving in a car, and imagine looking out the window. The side of the road seems to whiz by at high speeds, but distant mountains will change their apparent position slowly. Would you expect, then, that Rigil Kentaurus and Alpha2 Centauri are close to the Sun or very distant?

2. Look up the distances of these stars. For each star, run the cursor over the stellar image until it changes to the heads-up display shape (▤). Then right-click and select SHOW INFO. The distance is located under POSITION IN SPACE. If this has a + near it, click the plus to display the information. Write down the distance from the Sun for each star.

3. Do the same for three other bright stars near Rigil Kentaurus and Alpha2 Centauri. Write down their names and distances. Did you make the right guess in question 1?

ACTIVITY 2 – THE FASTEST-MOVING STAR

- Click the NOW button to reset the time. The flow of time will start, so hit ■ to stop it.
- Click on the FIND tab to open it.
  ○ Enter the name **Barnard's star** in the dialog box.

  ○ Double-click on its name to center the gaze on this star. As before, this should show the label of the star. If not, click on the box to the left of the star name.
- Click on the FIND tab to close it.
- Zoom all the way out using the mousewheel or the – button at the top right under the ZOOM display.

4. What constellation is Barnard's star currently in?

- Zoom in until the value of the height reads about 4°.
- Set the TIME FLOW RATE to 999 sidereal days.
- Start the flow of time. The view will be centered on Barnard's star, so it will seem to be stationary while the background stars move.
- Eventually, Barnard's star will enter a different constellation. You will notice this when the star crosses a line running across the display. Let the time run until the star is well within the new constellation. Stop the flow of time, then zoom out until you can read the constellation names.

5. What constellation will Barnard's star enter about 3,400 years from now?
6. Barnard's star has a higher apparent motion than does Rigil Kentaurus. Do you expect it is closer or farther away from the Sun than Rigil Kentaurus?

- Zoom in until you can see an image for Barnard's star. Select the star and right-click to drop down a menu. Select SHOW INFO.

7. What is the distance to Barnard's star?
8. What other factor besides distance determines a star's apparent motion?

Name: _____

Class/Section: _____

Starry Night Student Exercise – Answer Sheet
The Neighborhood of the Sun

1. Do you expect that Rigil Kentaurus and Alpha2 Centauri are nearby or distant stars?

_____

2. What are the distances from the Sun to Rigil Kentaurus and Alpha2 Centauri?

_____

3. What are names of three nearby stars and what are their distances? Did you make the right guess in question 1?

_____

_____

_____

_____

4. What constellation is Barnard's star currently in?

_____

5. What constellation will Barnard's star enter about 3,400 years from now?

_____

6. Barnard's star has a higher apparent motion than does Rigil Kentaurus. Do you expect it is closer or farther away from the Sun than Rigil Kentaurus?

_____

7. What is the distance to Barnard's star?

_____

8. What other factor besides distance determines a star's apparent motion?

_____

Beyond the Milky Way

GOALS

- To show that even nearby galaxies are located at vastly greater distances than the stars
- To examine the Milky Way and other galaxies

READING

- Section 20.1 – Galaxies Come in Many Types
- Chapter 22 – Modern Cosmology

One of the most difficult concepts to appreciate is that objects in the universe are located at tremendously large distances from the Sun. Our brains are not very good at imagining large numbers, though we can deal with them because we have convenient names such as "billion." In order to explore the size of the universe, we have to move away from the Sun in geometric steps that increase in distance by a multiplicative factor at each step. For example, if we doubled the distance at each step, we step out like this: 1, 2, 4, 8, 16, 32, . . .

In this exercise we will follow a similar pattern, only with much bigger steps. On the way outward from the Sun, we will first show that the stars are vastly farther away than any of the planets in the Solar System. Similarly, all the stars we see at night take up only a tiny part of our galaxy, the Milky Way. The galaxy is, in turn, only one of an immense number of galaxies that stretch out to very great distances. While going through the exercise, remember that the Starry Night program contains in its catalogs only a tiny fraction of all stars and galaxies.

SETUP

- Start Starry Night.
- Stop the flow of time with the Stop Time button (■).
- Click the Options tab to open.
  - Under Guides, turn all options off.
  - Under Local View, turn all options off.
  - Under Solar System, turn on Planets-Moons. All other options should be off.
  - Under Stars, turn on Stars. Leave other options off.
  - Under Constellations, turn all options off.
  - Under Deep Space, turn off all options.
- Click the Options tab again to close.
- Click the Find tab to open.
  - Under Solar System Items, click the boxes to the right of Earth, Jupiter, and Neptune. This will display the orbits of these planets.
  - Leave the Find tab open for now.
- Select Options / Viewing Location from the dropdown menus at the top of the Starry Night window. Click on the down-arrow (▼) next to the display box for View from. Select Stationary Location.
  - Under Cartesian Coordinates, replace the current numbers with

    X: 0 au
    Y: 0 au
    Z: 5 au

    Then hit the Go to Location button. The view will now shift so that you are looking down on the Sun and Earth's orbit from a position 5 au above the Sun. You should now see the complete orbit of Earth. Jupiter's orbit might be visible around the edges, and Neptune's orbit is outside your view.

- Click on the **V** tool to the left of the word Sun in the Find tab, and select Center from the dropdown menu.
- Close the Find tab by clicking on the word Find.

ACTIVITY 1 – STEPPING TO THE NEAREST STARS

- Using the procedure above, change your location so that you increase your distance from the Sun by a factor of 100. Under Cartesian Coordinates, enter
  - X: 0 au
  - Y: 0 au
  - Z: 500 au

1. The nearest star is located about 300,000 au away from the Sun. How many times farther away is the nearest star than your current location?
2. From this distance, is the orbit of Neptune easily visible? What about the orbit of Earth?

- Open the Find tab, and turn off the orbits of Earth, Jupiter, and Neptune.
- Close the Find tab.
- Now proceed outward by another factor of 100 in distance. Set your location to read
  - X: 0 au
  - Y: 0 au
  - Z: 50,000 au
- While the motion is taking place, look carefully to see if any stars seem to move. You may wish to set the location to Z = 500 au and again to Z = 50,000 au to notice any movement. The apparent movement would be caused by the change in viewing location. Only the nearest stars would move.
- If you did see movement, move the cursor over the star until it changes to the heads-up display shape (▤) and left-click to show its name. Then right-click and select Show Info. Write down the distance from the Sun, located under Position in Space. If there is a + located next to Position in Space, click on the + to show the data.

3. Which stars, if any, moved? What is their distance from the Sun?

ACTIVITY 2 – LEAVING THE MILKY WAY

- Click on the Find tab until it closes.
- Point at the star you selected, and right-click to see the dropdown menu. Click on Deselect (star name). The label for this star should disappear. Change your viewing location to

  - X: 0 au
  - Y: 0 au
  - Z: 80 ly

This is 100 times farther away than we were before, but we have shifted our units of distance to light-years. During the shift of position, many stars will appear to move.

- Just for fun, set the Time Flow Rate to 3000×. Start the flow of time with ▶, let it run for a while, then stop with ■. This will give you an impression that the stars are scattered through space in three dimensions.
- Now go out another factor of 100 in distance, so that the viewing location is
  - X: 0 au
  - Y: 0 au
  - Z: 8,000 ly
- At this point, we are located above the plane of the Milky Way. We can still see dots representing the stars near the Sun in the Starry Night catalog. Turn these off by opening the Options tab, and clicking the check box labeled Stars. Also turn off Planets-Moons.
- Under Deep Space turn on Tully 3D Database.
- Open the Find tab. In the dialog box at the top, type **Milky Way** and hit Return. The gaze will now shift so that you are aimed at the center of the Milky Way galaxy and can see the bright central bulge. A label for the Milky Way will also appear.
- Click the ⌃ button a few times under the Viewing Location display toward the top of the Starry Night window so you can see the image of the Milky Way. Out in deep space are several points representing distant galaxies.
- Start the flow of time with ▶, let it run for a while, then stop with ■. You will be tracing a big orbit around the Milky Way and can see its shape from different points of view.

4. Is the Milky Way a spiral galaxy or an elliptical galaxy? What features of the Milky Way tell you what type of galaxy it is?

- There are two satellite galaxies that are near the Milky Way. Use the Time Flow buttons to position these so you can identify them. Move the cursor over each image until it changes to the heads-up display shape (▤), then click to label them.

5. What are the names of these two satellite galaxies?

ACTIVITY 3 – DISTANT GALAXIES

- Remove the label from the last galaxy you marked by right-clicking on it and choosing Deselect (galaxy name).

- In the dialog box in the FIND tab, enter **Andromeda Galaxy** and hit ENTER. The view will now shift so you are looking at this large galaxy.
- Click the **V** button in the FIND tab next to the entry labeled Andromeda Galaxy (Spiral Sb Galaxy). Select GO THERE. The picture will zoom so that you are located closer to Andromeda.
- Start the flow of time, let it run for a while, then stop. While the time is running, pay attention to the arrangement of distant galaxies (all the dots).
- At Andromeda, we are located about 2 million ly from the Sun. Now go out another factor of 100 in distance, so that the viewing location is

    X: 0 au
    Y: 0 au
    Z: 200,000,000 ly

You will need to type in the commas or carefully count the zeros. As the distance shifts, the three-dimensional arrangement of distant galaxies is apparent. You may wish to see this by running the flow of time for a while.

6. Do the distant galaxies seem to be uniformly distributed in space or concentrated in groups?
7. What assumptions about the arrangements of galaxies do we make when studying the whole universe? Do you think we are far enough away from the Sun for these assumptions to be valid?
8. The Hubble constant is 22 km/second/Mly. Using Hubble's law, compute the recession velocity of a galaxy located at our current position (200 Mly from the Sun).